ASHFALL FOSSIL BEDS

State Historical Park & National Natural Landmark

Present View of an Ancient Past

A Cooperative Project of the University of Nebraska State Museum
and Nebraska Game and Parks Commission

Ashfall Fossil Beds State Historical Park | Royal, Nebraska | (402) 893-2000 | ashfall.unl.edu

By Mike Voorhies, Rick Otto, Sandy Mosel, Shane Tucker
Edited by Sandy Mosel, Rick Otto
Illustrations by Mark Marcuson, Angie Fox, Joel Nielsen, Adrienne Stroup
Layout and cover by Angie Fox
Photos by Rick Otto, Mark Harris of the University of Nebraska State Museum,
       Annie Griffiths/National Geographic Society, and
       NEBRASKAland Magazine/Nebraska Game and Parks Commission

UNIVERSITY OF NEBRASKA
STATE MUSEUM

N

NEBRASKA
— GAME PARKS —

Revised edition

©2015, 2016, 2022

Contents

Photo by Mark Harris, University of Nebraska State Museum

Introduction

A drive across northeast Nebraska on U.S. Highway 20 is pretty much what one might expect in the rural areas of the nation's heartland. The countryside consists of fields of corn and soybeans, and patches of small grains and alfalfa, interspersed with rolling pasture and grazing cattle. There is certainly nothing to indicate to the traveler that twelve million years prior one would have been journeying through a vast North American savanna, teeming with exotic wildlife. It takes some imagination to visualize herds of rhinos, delicate horses, camels and elephants grazing great seas of grass or browsing on shrubby trees where row crops now stand. Perhaps it seems even more unlikely that there could be much evidence of this ancient ecosystem anywhere in soil that has been tilled or pastured for over a century. But there is evidence—and it is abundant and graphic.

Located one hundred miles west of Sioux City, Iowa, and seven miles north of Highway 20 in northwest Antelope County, Nebraska, the Ashfall Fossil Beds State Historical Park has proof-positive that the subtropical savanna existed. Complete, articulated skeletons of several kinds of prehistoric mammals lay entombed there in a thick bed of volcanic ash. In all, over two hundred complete, or nearly complete skeletons have been exposed with the possibility of many more as excavation continues. The Ashfall site rests on a hill near a tributary of the Verdigre Creek, which flows into the Niobrara River thirty miles to the northeast. This river system, running the length of northern Nebraska, has been abundant in fossil mammal remains since the 1860's when paleontologists began collecting specimens in the area. The Ashfall site, however, remained unknown until 1971 when paleontologist Mike Voorhies noticed a thick layer of volcanic ash exposed by recent rainfall, located high on the side of a steep ravine. Closer investigation revealed an intact skull and lower jaw of a baby rhinoceros eroding out of the ash. Careful excavation yielded a virtually unheard of phenomenon—the entire rhino skeleton, and several more. With funding from the National Geographic Society, Dr. Voorhies was able to direct a major excavation of the site during the summers of 1978-79 and the results were astounding, with literally dozens of complete skeletons of rhinos, horses, camels and birds unearthed.

The most abundant skeletons at the Ashfall site, with over one hundred excavated to date, are of *Teleoceras major*, a short, stocky rhino with stout legs and a large barrel-like mid-section. With so many skeletons available from one site, it is possible to do a population analysis. Much about behavior and ecology that may not be evident in single animals can be determined from population studies. The population study at Ashfall suggests that the rhinos probably formed harems. There is a distinct absence of any number of young adult males and there is only one mature adult male for every five adult females. Numerous skeletons of young calves have also been found, many of them nestled next to adult females—almost certainly their mothers. The calves are generally of the same age, indicating a definite calving period. Like modern hippos, these animals probably spent time wallowing in waterholes, and they were grazers, with large, high-crowned teeth designed for a diet of grass. Fossil grass seeds are often found lodged between the teeth in the extinct rhino skulls. Close study of the fossil seeds reveals the grasses that grew on the ancient savanna were subtropical species, much like those currently found in Central America—an indication that the Great Plains' climate was most likely subtropical at the time.

Supporting this climatic evidence is the presence of fossil tortoises at the site. These large land reptiles cannot survive in temperatures that dip below the freezing mark. Giant tortoise fossils are reliable evidence that the climate in Nebraska was warmer twelve million years ago than it is now. There is also evidence that the climate was drying out. Nebraska fossils from the late Miocene indicate woodland environments were giving way to grasslands and the numerous horse skeletons at Ashfall are an excellent example. Some Ashfall species show varying degrees of tridactyly (three-toedness) in one population. This is evidence that three-toed species, at home in the forests, began to move onto the open grassland to take advantage of the availability of new niches. The harder ground of the open grassland reduced the usefulness of the side toes, which provided extra traction on the softer

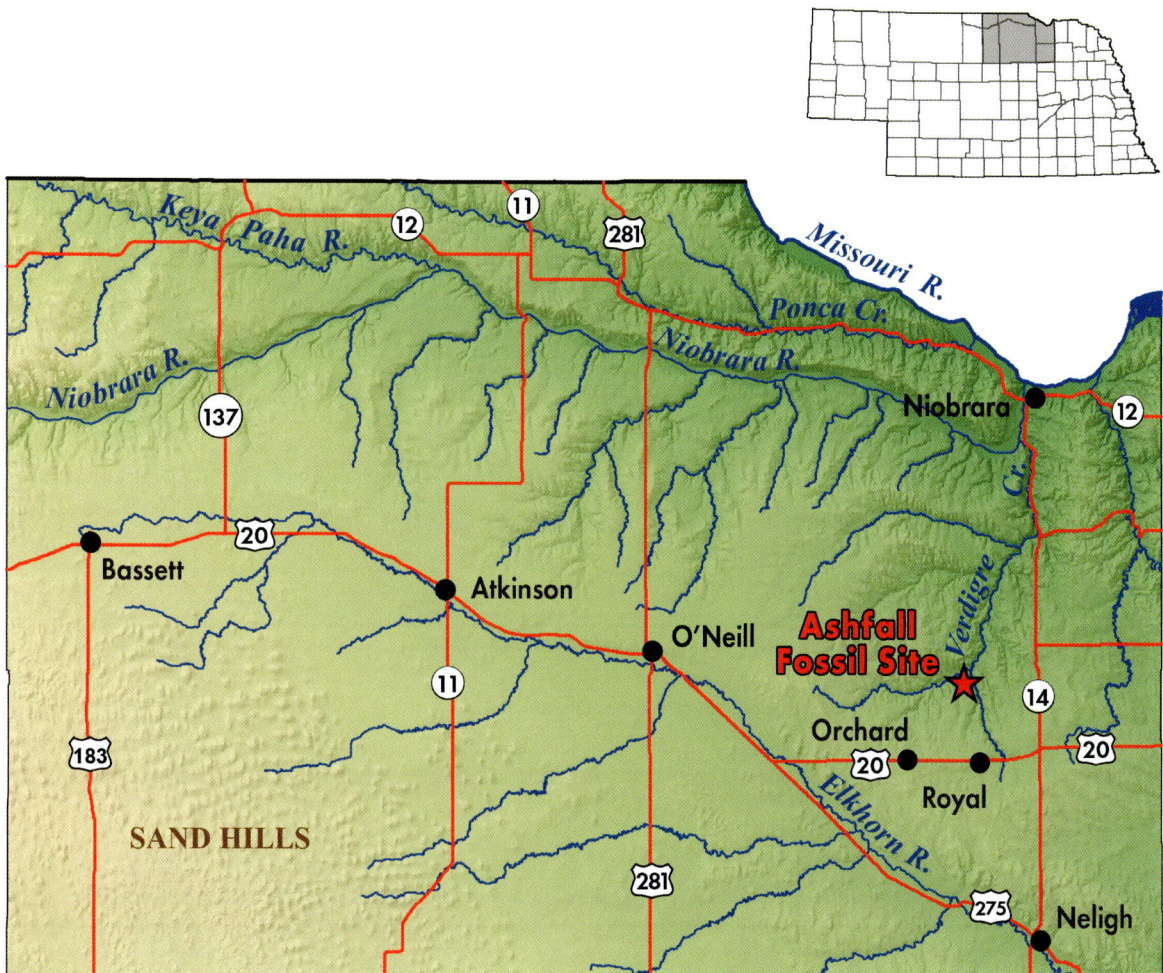

Map by Les Howard, University of Nebraska–Lincoln School of Natural Resources

forest floors, and the side toes began to reduce and disappear altogether. The horses at the Ashfall site clearly demonstrate evolution in action through the variances in their toes. It would not be possible to determine these changes as easily without the number of skeletons available for study at the site or the excellent condition of the delicate side toes, which are rarely preserved. Due to a relatively quick burial, however, complete preservation of skeletal remains is the rule rather than the exception at the Ashfall site. The vast number of skeletons and the quality of preservation at the Ashfall site are a vertebrate paleontologist's dream.

Research at the site has led geologists to believe that this mass graveyard is the result of an enormous volcanic eruption that occurred twelve million years ago in what is now southwestern Idaho, nearly nine hundred miles due west of the Ashfall Fossil Beds. Chemical analysis of the ash has determined its origin to be a massive volcanic caldera in the Bruneau-Jarbidge area—which produced an eruption one thousand times larger than the 1980 Mount St. Helens blast. The fine volcanic ash from an eruption of such magnitude would have been deposited hundreds of miles from the volcano. The prevailing westerly winds would have carried the ash eastward, blanketing a huge area of the Great Plains. The ash and other sediments from much of the last fourteen million years are still intact across most of northern Nebraska. Located within the Ash Hollow Formation, there are exposures of the ash layer in various locations along a 250-mile

Photo by NEBRASKAland Magazine/Nebraska Game and Parks Commission

stretch of the state. The ash layer is approximately a foot thick throughout. At the Ashfall site, however, it is eight to ten feet thick. Research has determined that the site was the location of a watering hole, basically a large depression on the landscape that would retain water during the wet season and dry out during dry spells, much like waterholes dotted across the savannas of Africa today. Fossils of pond turtles, remains of reed-like plant stems, and diatoms within the ash are evidence of water, but the lack of fish fossils and the absence of channeling in the strata indicate something less permanent than a lake or stream-fed pond. Nevertheless, the watering hole would have been a source of drinking water for local animals, as well as a place to wallow or bathe.

As thick clouds of the volcanic dust drifted in and settled on the landscape the animal community was devastated. The ash, which consists of tiny shards of glass, is extremely abrasive and dangerous when airborne in large quantities. Small animals, especially birds, would have succumbed almost immediately as the ash storm occurred. Larger animals would have survived the initial ashfall, but then encountered a landscape covered with a thick layer of fine, powdery, abrasive dust that buried their food supply and became airborne with every step or even the slightest of breezes. There was no escape and the horses, camels, and finally the large rhinos died of a slow suffocation as their lungs filled with volcanic ash in the days and weeks that followed. It seems likely that the animals came to the watering hole for some comfort, to lay in the coolness of whatever water was present as they became ill and feverish. As they perished, the constant winds across the open savanna caused the ash to drift, filling the waterhole where the animals lay, little by little until their carcasses were completely covered with a thick, dense layer of volcanic ash. Skeletons were exquisitely preserved in the round as carcasses rotted and ash took the place of muscle and organs. Young and old alike, skeletons of rhinos, horses, camels, deer, and birds were preserved in the time and place of their death.

In 1991, after twenty years of research, the Ashfall site was opened as a state historical park for the benefit of public education and enjoyment. At that time, a 2,000 square foot structure, aptly named the Rhino Barn, was built over a portion of the fossil bed so paleontologists could excavate fossils while visitors watched and inquired about the work taking place. The goal was to leave the fossils in situ, revealing more of the fossil bed each season, while leaving what had been previously excavated in place. In this way, the fossil bed gradually began to take shape. Excavation in the original Rhino Barn lasted ten years. At that time, there was no room left to work in the interior of the building, but there were more fossils yet to be revealed beyond its parameters. When funding became available in 2008 (via the Hubbard Family Foundation), a new Rhino Barn was constructed the following year. It covers 17,000 square feet and was built over the existing barn, which was then deconstructed, opening the fossil bed that had already been excavated to the areas yet to be exposed. The excavation resumed and continues each summer. Displaying the skeletons exactly as found creates a snapshot in time that visitors can observe and study. The on-going excavation showcases the delicate nature of the paleontological process in a rich but fragile fossil deposit. Video cameras catch the up-close action, which is displayed on monitors, allowing visitors a view as if they were sitting right where the excavation is taking place. The expanded Rhino Barn also contains a gallery of artwork depicting the animals that have been or might be found in the ashbed, and there is an exhibit explaining the source of the ash in detail. Additional exhibits will be added as time and funding allow. Displays and exhibits in the Visitor Center and along walkways and trails also aid in understanding the site. Something visitors can always depend on is a helpful park employee, ready to answer any questions that arise.

Excavation and research will continue at the Ashfall site for decades as long as funding is available. There is anticipation about what the next discovery may be. While much has been revealed, there are still many mysteries that remain. What about the elephants that would have been living at that time? What happened to the carnivores—saber cats and bone-crushing dogs that inhabited the area? Did they find a means of survival or are they waiting to be discovered in some as-yet unexcavated area of the fossil bed? What new species of plant or animal might yet lay buried in the ash that took their life but preserved their death?

A Detailed Look At Ashfall

The Ashfall Fossil Site (N42 25.255', W98 09.435') is located in northeast Nebraska approximately thirty miles east of the city of O'Neill, or, eight miles northwest of the village of Royal. The site is exposed on a highly eroded hillside above Colson Creek, an ephemeral tributary of the South Branch of the Verdigre Creek. The Verdigre Creek flows northwardly and joins the Niobrara River just a few miles from the Niobrara's confluence with the Missouri River. The Ashfall Site is situated near the southern margin of the lower Niobrara River Basin, approximately thirty miles south of the confluence of the Niobrara and Missouri Rivers. Over much of north-central Nebraska, the Niobrara flows eastward at a gradient of nine feet per mile; degradation during the Pleistocene and Holocene has eroded an east to west trench along the main river, with north-south headward eroding canyons more than two hundred feet deep in many places (as described by Skinner and Johnson, 1984). This region of northern Nebraska has been known for fossil remains eroding from unconsolidated sands since the earliest European explorers trekked onto the Great Plains.

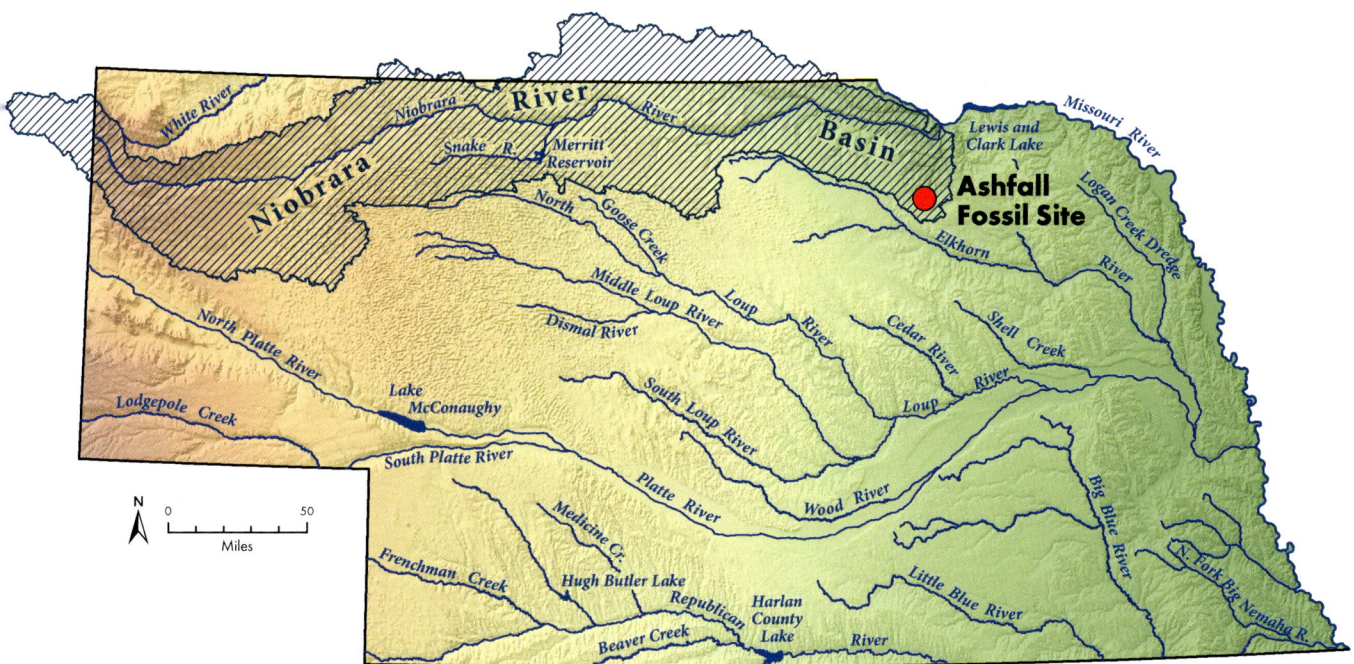

Map by Les Howard, University of Nebraska–Lincoln School of Natural Resources

Historical Context

Vertebrate fossil remains were reported from the Niobrara Valley as early as 1796 when Scotsman explorer James Mackay was mapping the middle Missouri River. Mackay described "the middle part of the thigh of an animal the large end of which was 7 inches in diameter and the other 6¾ inches." (Diller, 1955). This find may well be the "Ossemens de mamouth" (bone of a mammoth) labeled on a Mackay drafted map between the Keya Paha and Niobrara Rivers. (Nasatir, 1952). F.V. Hayden collected numerous specimens while serving as geologist for the Warren Military Expedition of 1857. This collection was sent to Joseph Leidy at the Philadelphia Academy of Natural Sciences who recognized its significance to paleontology in North America (Skinner and Johnson, 1984). By the turn of the 20th century, paleontology field expeditions from Yale Peabody Museum, American Museum of Natural History, University of Nebraska State Museum, and Amherst College had collected in northern Nebraska. In the following decades, the University of California Museum of Paleontology, University of Michigan Museum of Paleontology, Michigan State University, Notre Dame University, and National Museum of Natural History joined the field collections by sending dozens of prominent paleontologists to contribute to the research of Tertiary fossil vertebrates in the exposures along the Niobrara (Skinner and Johnson, 1984).

Even so, it was the dedication of one Nebraska-born paleontologist whose extensive contributions set the foundation for the understanding of the geology and paleontology of the Niobrara River Valley. In 1926, Morris Skinner began a career of extensive collecting for the Childs Frick Lab at the American Museum of Natural History. His active fieldwork spanned into the 1980s and his great contributions were the attention to detail of stratigraphic sequences and the recording of fossil remains found in specific horizons of the strata. His documentation of the biostratigraphy of the Niobrara Valley resulted in a more thorough understanding of how the Great Plains have transformed over the past fifteen to twenty million years. Skinner's six major research publications have produced a comprehensive record of the transition of species over time, climatic change, depositional environments, and physiographic transformation during the Miocene, Pliocene, and Pleistocene epochs.

Mike Voorhies excavates rhino fossils in 1987.

History of the Locality

Morris F. Skinner was a mentor to Mike Voorhies during his professional career. Like Morris, Mike grew up in a small town in northern Nebraska. Mike's hometown of Orchard was within walking distance of the tributaries of Verdigre Creek where locals collected fossil bones and petrified wood eroding out of the ground. In his childhood, Mike's interest in nature also led him to a curiosity of fossil bones. During his education, it led him to geology and vertebrate paleontology.

While attending the University of Nebraska, Mike was able to take part in field research and excavations, including the recovery of a remarkable find; two bull mammoth skeletons, with tusks interlocked, near the panhandle town of Crawford. (The Crawford mammoths are now on display at the Trailside Museum of Natural History at Fort Robinson State Park near Crawford.) Mike Voorhies' graduate training took him to the University of Wyoming where he studied under Paul O. McGrew. Mike's graduate research was based on a fourteen-million-year-old stream channel deposit not many miles from his hometown. The dissertation is titled *Taphonomy and Population Dynamics of an Early Pliocene Vertebrate Fauna, Knox County, Nebraska*, Contributions to Geology, Special Paper No. 1, University of Wyoming, July 30, 1969.

After a few years on faculty at the University of Georgia, Mike was conducting field excursions back in the Verdigre Creek Valley of Nebraska, mapping exposures of Miocene and Pliocene strata, and correlating them with the exposures Morris Skinner had documented one hundred miles to the west. It was on one such day of mapping in May of 1971 that Mike noticed a small rhino jaw protruding from the side of a gully on the Melvin Colson farm in northern Antelope County. The lower mandible was articulated with the skull, and the fossil was embedded in a pure glassy volcanic ash. A few more minutes of probing and brushing and he realized that the skull aligned with the cervical vertebrae. Another hour of excavation provided the realization that the rest of the skeletal elements were in place, and a fully articulated fossil rhino calf was embedded on the precipice of a nearly vertical embankment. This find was in close proximity to where University of Nebraska State Museum paleontologists Lloyd Tanner and Henry Reider collected an articulated skull and mandible from an adult *Teleoceras major* in 1953 (Holt County Frontier, October 8, 1953). It was several years before Voorhies and a few graduate students returned to the site. A more in-depth survey in 1977 provided evidence of multiple skeletons buried in the volcanic ash, and it was at this point that a successful grant proposal was submitted to the National Geographic Society for funding of a major excavation of the locality named "Poison Ivy Quarry".

The discovery of one small rhino jaw is the beginning of what is now the Ashfall Fossil Beds State Historical Park.

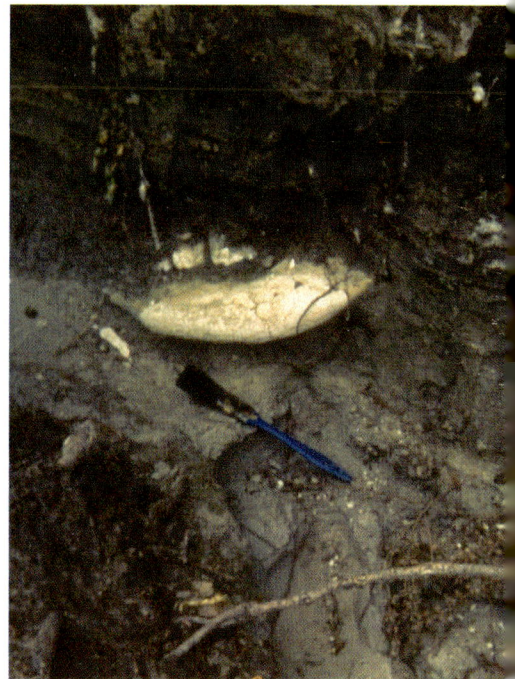

Above: Photo by Mike Voorhies

Facing page: Photo by NEBRASKAland Magazine/ Nebraska Game and Parks Commission

Photo by NEBRASKAland Magazine/Nebraska Game and Parks Commission

Excavation of initial test pits indicated that the skeletons were limited to the lower three feet of a ten foot thick bed of pure volcanic ash. Access to the skeleton level of the ashbed was complicated by ten to thirteen feet of sand and soil superimposed on the ashbed. The 1978 excavation season began with the mechanical removal—by bulldozer—of the overburden. Diligent scouting by the field crew also allowed for removal of the upper portion of the volcanic tuff in an area of approximately three thousand square feet. What was revealed in the following weeks had not been observed anywhere before: Dozens of three-dimensionally preserved, articulated carcasses, laying side-by-side. Another pattern was developing as well; the barrel-bodied rhinos (*Teleoceras major*) were situated above skeletons of smaller size, such as the horses (*Pliohippus pernix, Protohippus simus, Cormohipparion occidentale, Neohipparion affine, Pseudhipparion gratum*) and camels (*Procamelus grandis, Protolabis*). Lower yet in the ashbed strata were the smallest size species, examples include the long-snouted saber-tooth deer (*Longirostromeryx wells*i), several species of birds; crowned cranes (*Balearica exigua*), long-legged hawk (*Apatosagittarius terrenus*), an old world vulture (*Anchigyps voorhiesi*); and two species of water turtle (*Chrysemy*s, a new species and *Sternorthurus odoratus*,

An aerial view of a portion of the Ashfall Fossil Beds site. This photo was taken during the Park's development and shows a series of six-meter sampling grids layed out on the newly-exposed surface of the ashbed. The overburden (sand and sandstone) and the upper three to five feet of volcanic ash have been removed, leaving the fossils safely buried under another five feet of ash. The Rhino Barn (not yet built when this photo was taken in 1989) now covers the area at the right side of the picture. Exploration has indicated that this is a very small portion of the actual bone-bed.

a musk turtle). It became apparent to Voorhies that the smallest-sized species in the fauna perished before the larger-sized species. Extensive removal of volcanic ash from the surface of the sand substrate revealed evidence of trace fossils including ungulate tracks that filled with the initial settling of volcanic ash. Contours in the surface of the exposed sand substrate revealed that the ash settled into a sizable depression on the landscape. The stratified ash, lateral to and above the carcasses, suggest that airborne ash had settled into a pool of water without current flow; shallow water ripples, and aquatic animal and plant species, also indicate the presence of water in the depression at the time of the animals' death.

Two seasons of excavation in 1978 and 1979 by students, staff, and volunteers from the University of Nebraska State Museum recovered a significant sample of specimens from the site. During the hot summer days, curious local residents would walk a half-a-mile or more to sit above the quarry and watch the progress of the dig. One of the most interesting aspects of the fossil bed was the graphic quality of the full-shaped skeletons that impressed the onlookers. Inspired by the public interest and visual quality of the in situ specimens, Mike Voorhies began to plan the development of the site into a working excavation accessible to anyone who wanted to observe the process of field work in vertebrate paleontology.

After six years of planning and negotiating between two state agencies, the University of Nebraska–Lincoln and the Nebraska Game and Parks Commission, a workable plan came together. Exploratory excavation of the ashbed in the late 1980s revealed that the deposit was quite extensive, and that a protective enclosure for the fossil bed, dubbed the Rhino Barn,

Ashfall Fossil Beds was originally known as the "Poison Ivy Quarry." The first major excavation at the Ashfall site was conducted in the late 1970s before development as a State Park. All fossils were removed for research and safe-keeping during this National Geographic Society sponsored dig.

Photo by NEBRASKAland Magazine/Nebraska Game and Parks Commission

The original Rhino Barn provided ten seasons of excavation in a two thousand square foot area. Thirty-three skeletons were revealed in this space. This "Barn" was replaced by the Hubbard Rhino Barn in 2009.

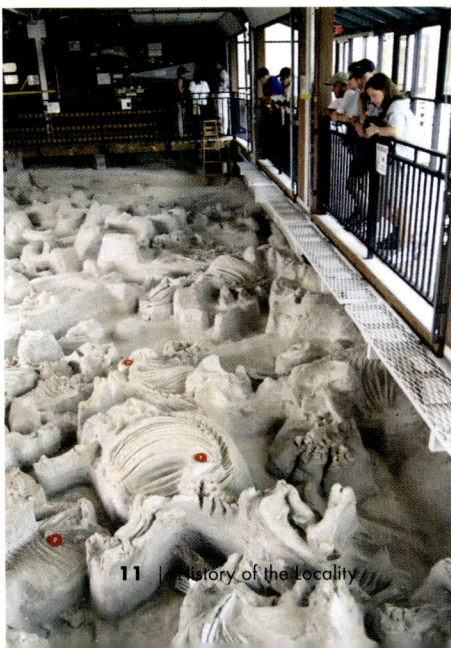

was justifiable. After private donations of $385,000 were secured for land purchase and construction of roads and structures, the Ashfall Fossil Beds State Historical Park opened to the public on June 1, 1991. Within days of opening Ashfall Park, barrel-bodied rhino skeletons were in the process of being exposed and visitors from across the United States swarmed to the Ashfall Fossil Site to view the activity—and fossil skeletons—in the newly opened Rhino Barn.

Excavations are conducted by college students, hired as interns, who spend twelve weeks during the summer conducting the dig. After ten seasons of excavation, the area inside the two thousand square foot Rhino Barn was entirely revealed. All fossils were left in situ within the confines of the Barn. Several years passed with little excavation in the volcanic ashbed, but shortly after the site was designated a National Natural Landmark in 2006, a philanthropic family from Omaha pledged to fund a new fossil bed enclosure 17,500 square feet in size. The new Hubbard Family Foundation Rhino Barn opened in 2009, and ashbed excavations have resumed again. As of publication, fifty-two articulated skeletons representing nine of the eighteen known ashbed species are exposed in a four thousand square foot section of the Barn.

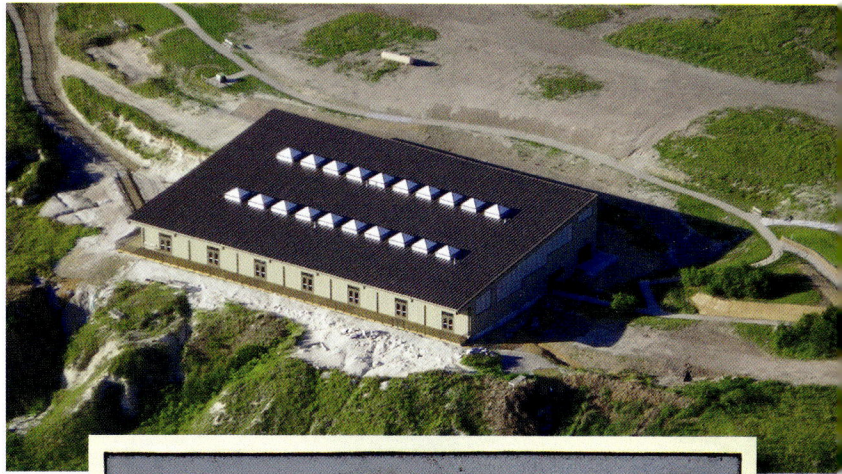

The 17,500 square foot Hubbard Rhino Barn will provide many seasons of excavation and discovery. All fossil skeletons will remain in situ (in place) as they are exposed by student paleontologists.

The exceptional preservation of the fossil skeletons in the ashbed prompted the National Park Service to declare the site a National Natural Landmark in 2006. The site was deemed to be a prime example of a specific geological feature, based on the clearly observable, detailed preservation of the fossils.

NEBRASKAland Magazine/Nebraska Game and Parks Commission

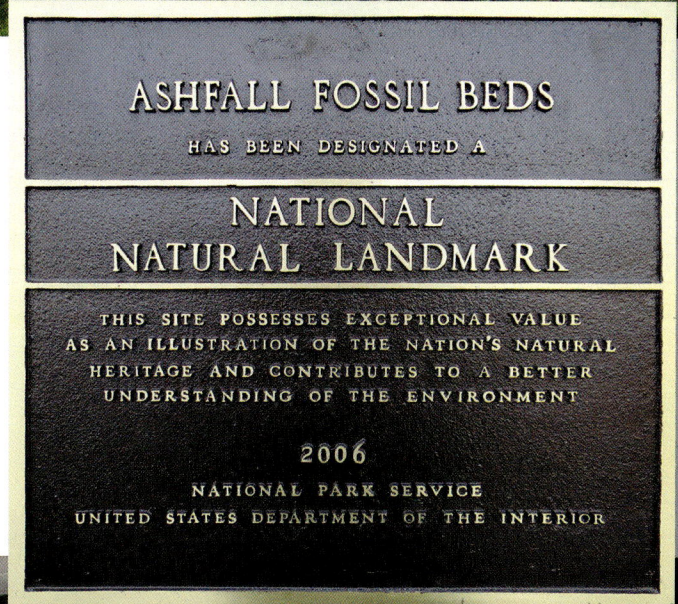

ASHFALL FOSSIL BEDS

HAS BEEN DESIGNATED A

NATIONAL
NATURAL LANDMARK

THIS SITE POSSESSES EXCEPTIONAL VALUE
AS AN ILLUSTRATION OF THE NATION'S NATURAL
HERITAGE AND CONTRIBUTES TO A BETTER
UNDERSTANDING OF THE ENVIRONMENT

2006

NATIONAL PARK SERVICE
UNITED STATES DEPARTMENT OF THE INTERIOR

The Hubbard Rhino Barn is constructed on part of an ancient waterhole that drifted full of volcanic ash almost twelve million years ago. The waterhole contains dozens of intact skeletons of animals that perished in the waterhole and were subsequently buried by volcanic ash.

EARTH'S CRUST SHIFTED WESTWARD OVER MILLIONS OF YEARS

NEXT ERUPTION
? ? ?

SNAKE RIVER PLAIN VOLCANIC PROVINCE

YELLOWSTONE NATIONAL PARK

SNAKE RIVER

Yellowstone Plateau Volcanic Field, active 2.1–0.6 million years ago.

Heise Volcanic Field, active 6.5–4.3 million years ago.

Picabo Volcanic Field, active about 8 million years ago.

Twin Falls Volcanic Field, active about 10 million years ago.

Bruneau-Jarbidge Volcanic Field, active 12.5–10 million years ago. Ashfall Park's ash originated here.

Owyhee-Humboldt Volcanic Field, active 13.9–12.8 million years ago.

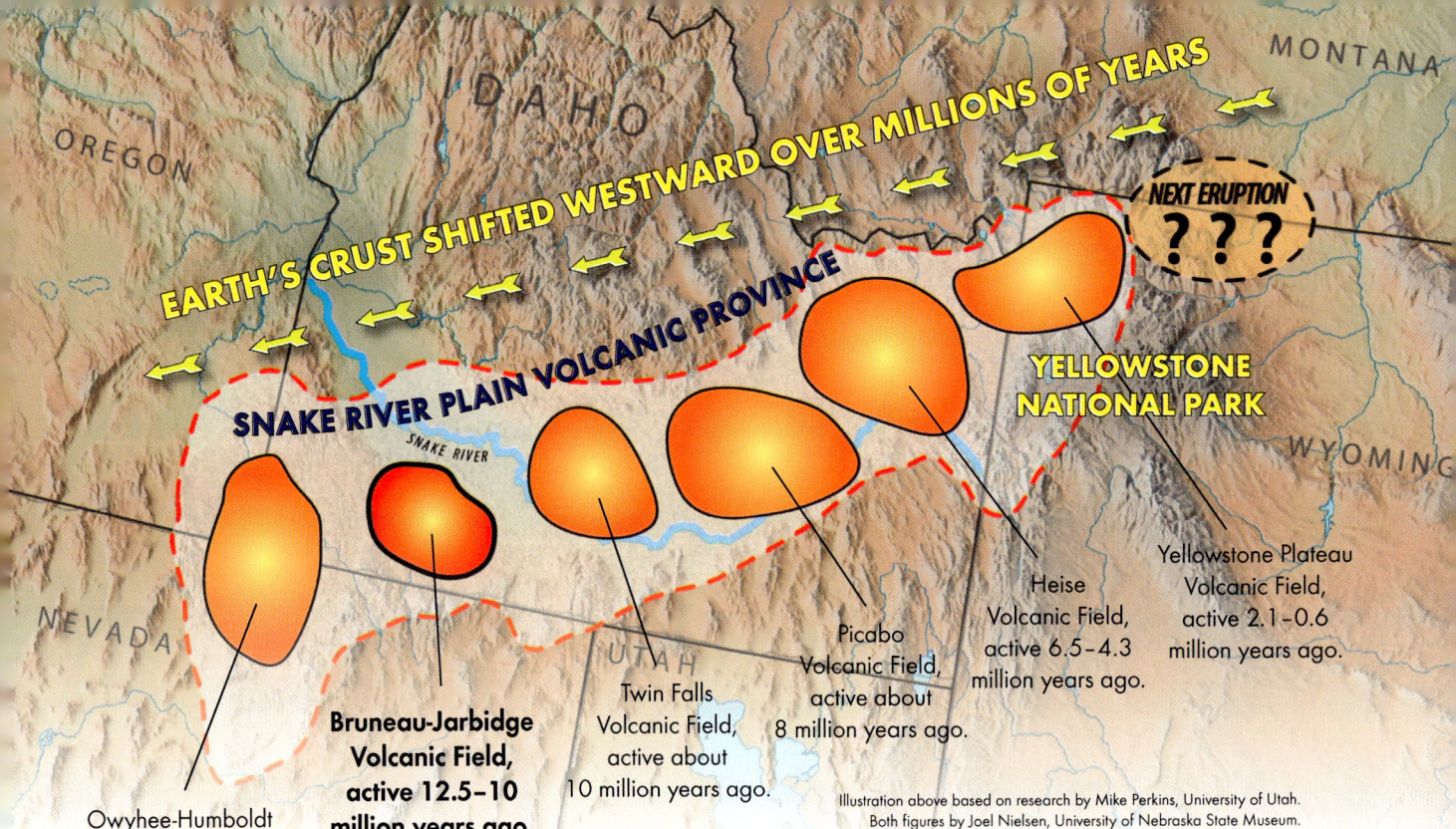

Illustration above based on research by Mike Perkins, University of Utah. Both figures by Joel Nielsen, University of Nebraska State Museum.

THE BRUNEAU-JARBIDGE SUPERVOLCANO

(Above) The Ashfall ash came from a gigantic explosive volcano in southwestern Idaho, the remains of which are called the Bruneau-Jarbidge Caldera. Ordinary volcanoes are conical mountains built of lava and cinders, but truly enormous volcanic eruptions don't create mountains, they destroy them. Following each huge eruption (about every five hundred thousand years in the western United States) the earth's crust collapses into the empty volcanic chamber, forming a wide, shallow crater called a *caldera*. The Bruneau-Jarbidge Caldera is over fifty miles in diameter. The Yellowstone Caldera is thirty miles across.

(Below) This giant crater south of the Snake River is the source of the Ashfall ashbed. Distribution and thickness of ash from Mount St. Helens (May 18, 1980) is shown for comparison.

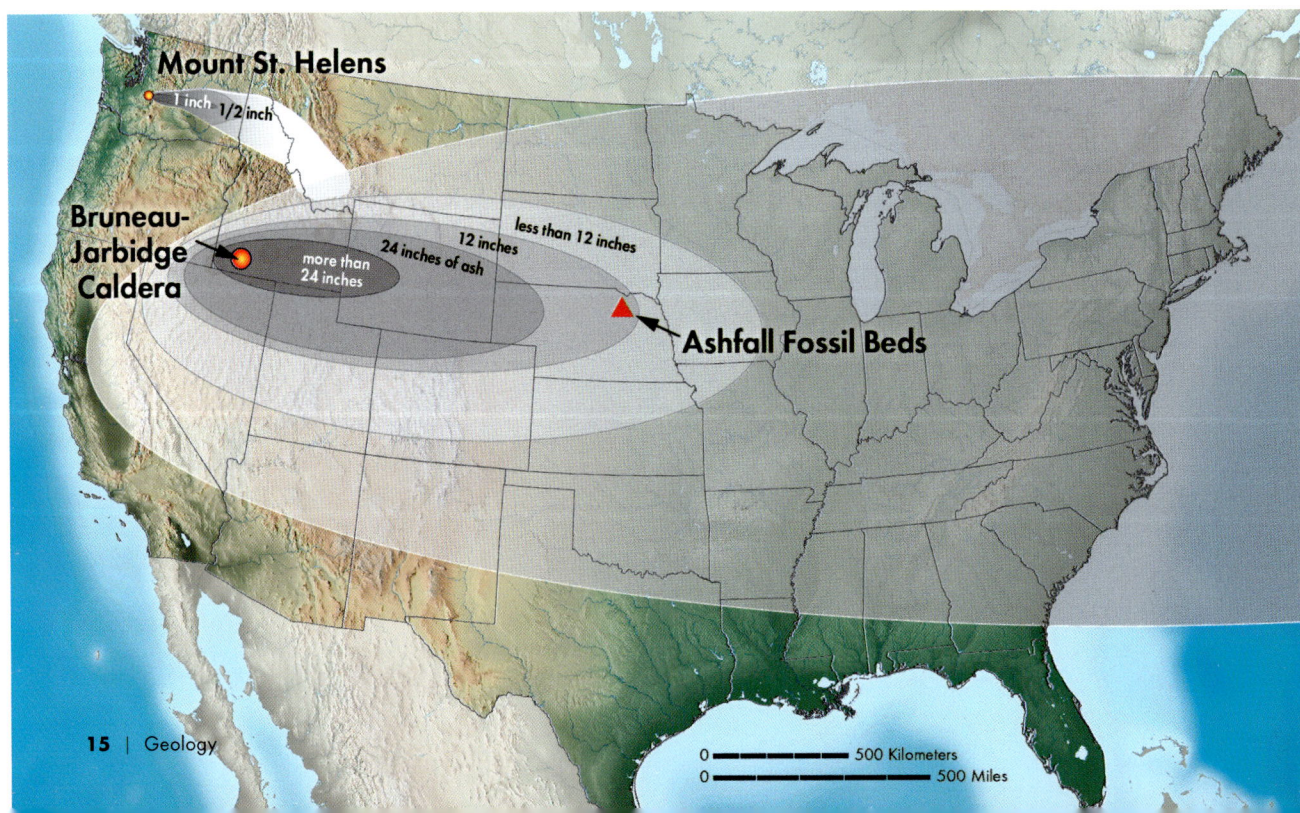

Mount St. Helens
1 inch   1/2 inch

Bruneau-Jarbidge Caldera

more than 24 inches
24 inches of ash
12 inches
less than 12 inches

Ashfall Fossil Beds

0 — 500 Kilometers
0 — 500 Miles

Fossil Bed Geology and Age

Volcanic Ashbed

The volcanic ashbed is eight feet thick where the skeletons are located in the ancient waterhole. The chemical composition of the ash provides a "fingerprint" of the source, and the age of the eruption. A quality of volcanic ash is that it contains radioactive isotopes which can be used to establish the date of a volcanic eruption. Several dating methods using radioactive isotopes have been developed. Fission-tracks (tiny scratches in the volcanic glass shards) from decay of Uranium was the first method used to date the Ashfall ash producing dates of 10.3–10.6 million years ago (MYA). With the advent of single-crystal laser-fusion Argon 40/Argon 39 dating, the source tuff of the Ashfall ash was determined to be 11.81 MYA[2]. Recalibration of volcanic ashbed dates to Fish Canyon Reference Standards[3] adjust the date to 11.93 MYA[4].

(Above) Volcanic ash, composed of small jagged glass shards, magnified 525 times.

When using radioactivity to date volcanic ash, zircons are the standard. Zircons are tiny crystals that form during a volcanic eruption and contain a radioactive Uranium isotope. The ratio of Uranium to Lead in the zircon is used to date the volcanic ash. Several attempts to recover zircons from Ashfall were unsuccessful, as the process literally compares to finding a needle in a haystack. In 2016, geologists from the University of Kansas successfully recovered zircons that were dated at 11.86 MYA[6], almost identical to the Argon40/Argon39 date.

Bonnichsen (1980) and Perkins, et al. (1995) independently determined the source of the Ashfall ash as the Bruneau-Jarbidge Caldera in southwest Idaho. One of the Yellowstone Hotspot Eruptive Centers, the Bruneau-Jarbidge was active from 10 MYA to 12.5 MYA. The solid volcanic rock from the 11.93 MYA eruption near the caldera matches in chemical composition with the Ashfall ash[5]. Geologists refer to the ash from this specific eruption as the "Tuff of the Ibex Hollow."

Strata Below The Ashbed

A silty sand substrate three to five feet thick underlies the volcanic ashbed. Referred to as the "waterhole sand," this unit contains an abundance of fragmented vertebrate fossils from animals that perished from illness, predation, or old age prior to the volcanic ash event. Some of the fragmented skeletal elements exhibit tiny striations on the surface that indicate trampling by large herbivores; while others have spiral fractures, scalloped edges, and punctures, the result of scavenging by one of several species of bone-crushing dogs from the now extinct family Borophaginae. The Ashfall waterhole sand is a fossil zone that correlates well with the "Fragmental Layer" in central and western Nebraska proposed by Walker Johnson in 1936.[7]

A "cap rock" caliche bed underlays the waterhole sand and is three to five feet below the base of the volcanic ashbed. The caliche bed is a calcium cemented sandstone, suggestive of a seasonally arid environment during the time of the Ashfall event.

1 Boellstorff, 1980
2 Perkins, et al., 1995
3 Cebula, et al., 1986, Lanphere and Baadsgard, 2001, Renne, et al., 1994, 1998
4 Perkins and Nash, 2002
5 Perkins, et al., 1995
6 Smith, Turner, Moller, et al., 2018
7 Also see Herbel, 1994

Cross-section of the waterhole filled with ash.

Cross–Section of Exposed Geology

BROADWATER FORMATION
(formerly the Long Pine Formation)
Age: 2–3 million years
Descripton: colorful Rocky Mountain gravel and sand
Common Fossils: teeth and bones of zebras, giant camels, *Stegomastodon*, beavers, muskrats

ASH HOLLOW FORMATION
(Cap Rock Member)
Age: 10–12 million years (ashbed is 11.9 million years)
Description: ledgy tan sandstone with gray ashbed
Common Fossils: teeth and bones of rhinos, three-toed horses, camels, four-tusker elephants, tortoises; whole skeletons of rhinos, camels and three-toed horses in ashbed.

VALENTINE FORMATION
(Devil's Gulch Member)
Age: 12–13 million years
Description: greenish silt and clay with limy nodules
Common Fossils: Weathered bones from rhinos, four-tusker elephants and three-toed horses

VALENTINE FORMATION
(Crookston Bridge Member)
Age: 13–14 million years
Description: tan river sand
Common Fossils: fossil wood, bones and teeth of tiny pronghorns, four-tusker elephants, rhinos, fish

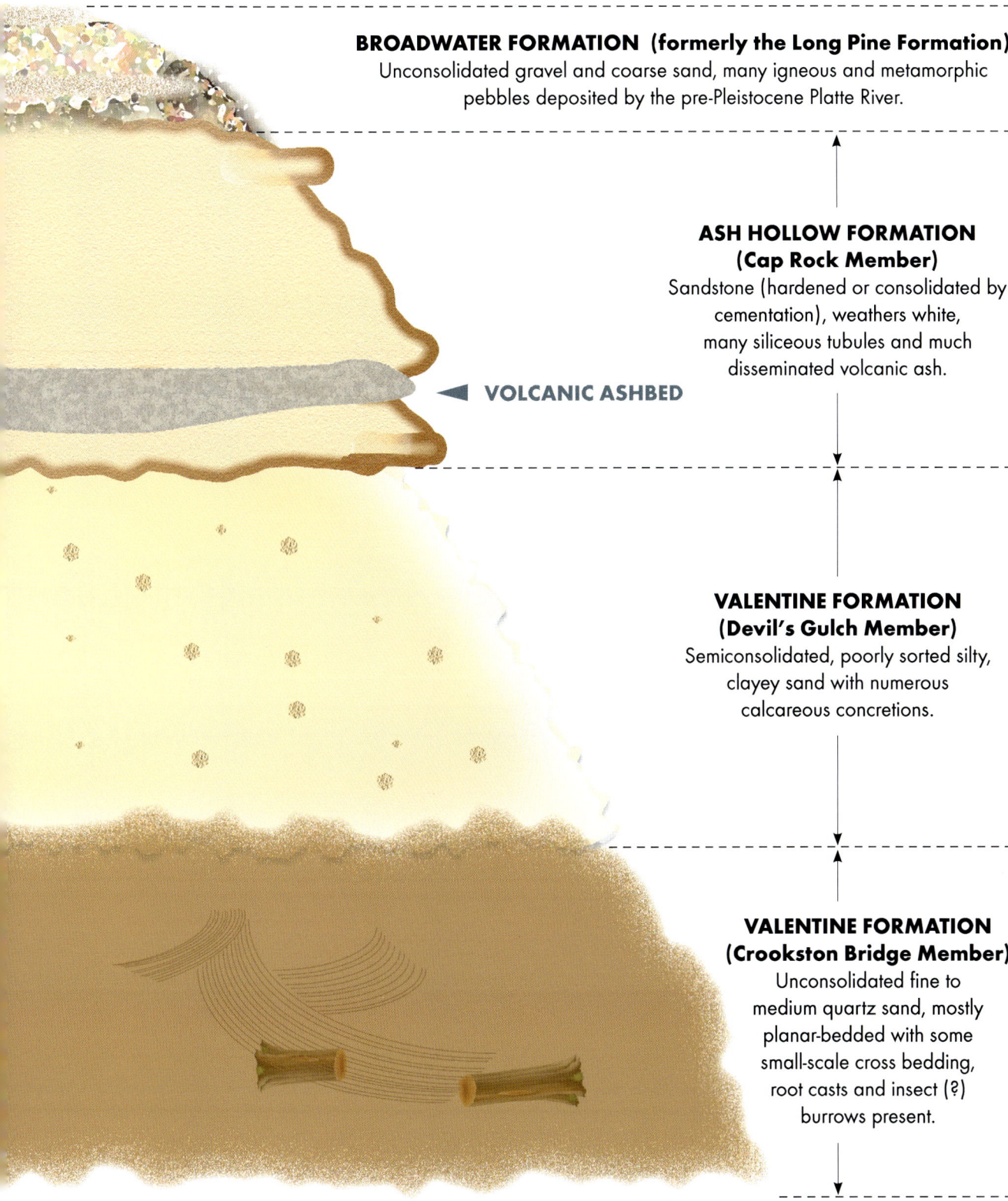

BROADWATER FORMATION (formerly the Long Pine Formation)
Unconsolidated gravel and coarse sand, many igneous and metamorphic pebbles deposited by the pre-Pleistocene Platte River.

**ASH HOLLOW FORMATION
(Cap Rock Member)**
Sandstone (hardened or consolidated by cementation), weathers white, many siliceous tubules and much disseminated volcanic ash.

◀ **VOLCANIC ASHBED**

**VALENTINE FORMATION
(Devil's Gulch Member)**
Semiconsolidated, poorly sorted silty, clayey sand with numerous calcareous concretions.

**VALENTINE FORMATION
(Crookston Bridge Member)**
Unconsolidated fine to medium quartz sand, mostly planar-bedded with some small-scale cross bedding, root casts and insect (?) burrows present.

How Old Are the Fossils?

Using a recently developed method of radiometric dating (single crystal laser-fusion, Argon40/Argon39), scientists date the Ashfall Fossil Beds at almost twelve million years old. That's much later than the dinosaurs but a long time before the Ice Age. Nebraska was already well above sea level by then but we still had a nice mild climate—no brutally cold winters or broiling summers.

Just how long is twelve million years? Imagine shrinking the entire four and a half billion year history of planet Earth down to the length of a 24-hour day. Under this scheme, Ashfall didn't happen until four minutes before midnight. From this perspective then, Ashfall happened almost yesterday. However, if we compare Ashfall's date with some familiar landmarks of human civilization—Ashfall is three thousand times older than the pyramids, and sixty thousand times older than Washington, DC—the fossil beds seem respectably ancient indeed. Scale is important!

Ashfall Fossil Beds at almost twelve million years old

Today

U.S. Department of the Interior/U.S. Geological Survey

How Reliable Are Radiometric Dates on Rocks and Fossils?

A good clock keeps ticking at the same rate and keeps track of the number of ticks no matter what happens. Before the 20th century, geologists were unaware that many rocks, especially ones formed during volcanic eruptions, do indeed contain tiny "clocks" (crystals) that can accurately be "read" with the proper tools. In order to be useful for dating, a crystal must contain an unstable isotope, a form of a chemical element constantly undergoing change (radioactive decay). It is such naturally-occurring isotopes that are the basis for determining the age (in years) of rocks including the Ashfall ashbed.

In Nature

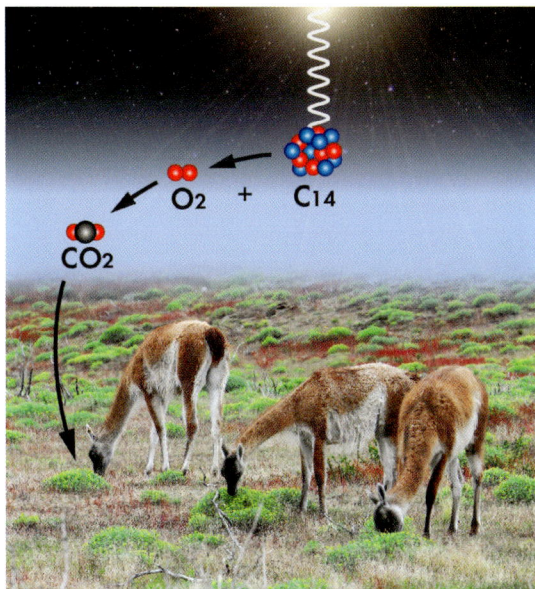

In the Laboratory

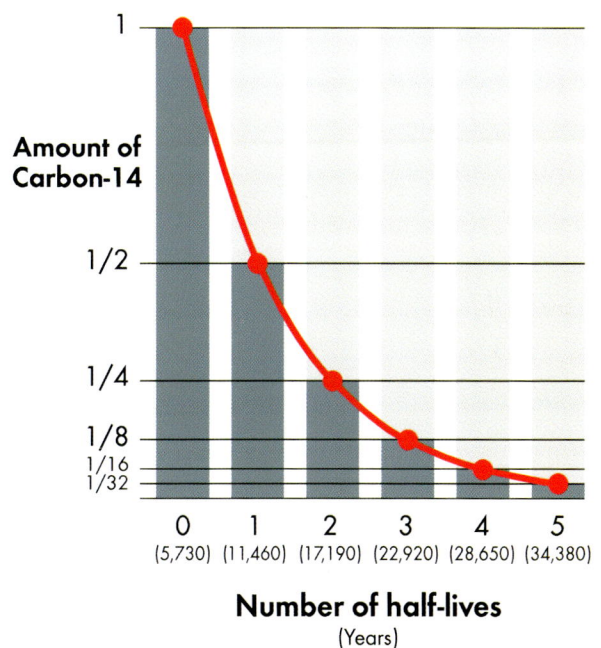

Number of half-lives
(Years)

Carbon-14 atoms are created by cosmic radiation in the upper atmosphere.

Carbon-14 combines with oxygen (O_2) to form carbon dioxide (CO_2) which is taken into the the bodies of all living things.

All living things have a small but stable amount of carbon-14 in their tissues but after death carbon-14 is slowly lost by radioactive decay, leaving behind only stable (nonradioactive) carbon-12.

The accuracy of radiocarbon dating has been demonstrated repeatedly by analyzing materials of known historical age (Egyptian artifacts up to 3,000 years old; tree rings up to 10,000 years old).

By measuring the amount of carbon-14 in a sample of wood, bone, or other organic material, we can tell the age of a sample.

Exactly half of the carbon-14 in a sample decays every 5,730 years (the "half life" of carbon-14). After 11,460 years, only 1/4 of the original C-14 remains, after 17,190 years the amount decreases to 1/8, etc. After about 10 half-lives, the amount of radiocarbon becomes essentially too small to measure. This means that material older than 50-60,000 years old can't be reliably dated by radiocarbon analysis.

For older material, geologists analyze isotopes with much longer half-lives like potassium, argon, and uranium.

Environment of the Volcanic Ash Deposit at Ashfall

Voorhies' 1985 report to the National Geographic Society (revised), interprets the details pertaining to the "Origin of the Deposit" as follows:

• The site was originally a shallow depression, possibly a pond formed as an oxbow, on the floodplain of a low-gradient stream.

Evidence: The ashbed is bowl-shaped, with thin continuous laminated layers of ash, and ripples within the layers indicating accumulation in standing water with little or no current action. The sandstone beneath the ashbed, in contrast, is made up of sand grains of varying sizes and fossil plant roots (rhizoliths). Some sand bedding indicates stream current.

• The site is a waterhole for local populations of large mammals.

Evidence: Numerous fragmentary bones and teeth (mostly rhinos, horses, camels, mastodonts, and turtles) occur in the silty sand just below the ashbed. These remains are broken, abraded and scattered, possibly by trampling.

• A volcanic ashfall buried the local landscape, probably to a depth of eight to twelve inches.

Evidence: This is the thickness of the lowermost portion of the ashbed at Poison Ivy Quarry; it is uncontaminated with sand and silt and therefore almost certainly represents direct air-fall ash.

• Large numbers of pond turtles and birds perished during the early phases of the initial ashfall. Herds of three-toed horses, camels and small deer died catastrophically during the late phases of the initial ashfall.

Evidence: Skeletons of the pond turtles and birds occur above the basal ash considered to be original air-fall material. Horses, camels and small deer are at a slightly higher stratigraphic level. Evidence of close succession of death is twofold: study of individual age at death of the animals and similar states of preservation of the animals.

• Some carcasses decomposed slightly and were subjected to scattering by trampling and scavenging.

Birds and pond turtles died early in the ashfall event, possibly within hours.

Evidence: The skeletons of the smaller animals are somewhat disarticulated and scattered. At least one horse skull shows two punctures almost certainly caused by the canine teeth of a scavenger. Many of the skeletons exhibit differential smashing (some parts well preserved, adjacent portions flattened and broken) in a manner highly suggestive of trampling by animals.

• Herds of rhinos and a few late-coming horses and camels arrived at the waterhole and died, again very rapidly, perhaps a week or so after the principal die-off of smaller animals.

Evidence: Almost all of the rhinoceros skeletons are lying on, not in, the direct air-fall ash. They frequently overlie smaller skeletons such as horses, camels, and birds, and are better articulated and less scattered. Studies of individual age indicates rapid death of the rhinos.

• The rhino skeletons were rapidly buried by ash, probably in a few weeks or less.

Evidence: The three-dimensional preservation of most of the rhinoceros skeletons, many of them in crouched positions with the legs directly beneath the body, indicates that the carcasses were still largely intact when buried.

• The ancient waterhole completely filled with ash, sealing in the skeletons. This was probably accomplished within a few weeks or less by wind blowing ash from the surrounding ash-blanketed countryside.

Evidence: Wind rather than water transport is suggested by the lack of sand and silt in the upper portions of the ash. Any significant erosion by water would surely have cut through the thin loose mantle of ash on the hillsides and washed sandy debris into the pond.

In summary, the fossils represent populations of animals that died catastrophically around a shallow waterhole during or shortly after a major ashfall that blanketed the surrounding countryside with volcanic ash. Smaller animals like horses, camels, and birds died first and began to decompose. Before disarticulation of the smaller animals was well advanced, but after deposition by ash, a large number of rhinoceroses entered the waterhole, trampling many of the smaller skeletons; the rhinoceroses eventually died and were covered by volcanic ash, which probably sifted into the waterhole as the ash blanket on the landscape was reworked by wind.

Horses and camels died after days of enduring the ash storm. Rhinos died some days or even weeks after the horses and camels.

Illustrations by Adrienne Stroup

Student paleontologist Maria Brown brushes the volcanic dust from a rhino skeleton. The three-dimensional preservation and close proximity of the skeletons to each other provides a very graphic image.

Photo by NEBRASKAland Magazine/ Nebraska Game and Parks Commission

How Did the Animals Die and How Long Did It Take?

It is apparent that the birds died very shortly after the ash dust began to fall from the sky since the bird skeletons are located in the lowest (basal) layers of the ashbed. As the ash accumulated on the sandy terrain, it may have taken only hours for the cranes, hawks and vultures to die. But the story seems quite different for the larger mammals. Many bones of the camels, horses, and rhinos have patches of rough, frothy bone growth on the surface of the normal bone. The pattern of accumulation of abnormal bone is associated with lung damage first described by French pathologist Pierre Marie and his colleague Eugen von Bamberger in 1890. Also known as Marie's Disease, or Bamberger-Marie Disease, the currently accepted technical term for the malady is *hypertrophic osteopathy*. It has been documented in humans and a variety of domesticated species. The abnormal bone growth is the secondary result of various organ diseases, but occurs frequently with lung-damaging diseases including tuberculosis and pneumonia, as well as tumors and abscesses. Symptoms begin with soft tissue swelling of the feet, fever, cough, and lameness. Left untreated, the thick pathologic bone growth develops on the limb bones and may spread to other parts of the skeleton.

There was no place for the local wildlife to escape the dense clouds of volcanic ash, or the inevitable inhalation of dangerous amounts of the ash when the camels, horses and rhinos were grazing—or attempting to graze on the glass-dust covered landscape. The rhinos lived longer than the smaller-sized species such as the horses and camels, and the animals lingered for weeks before succumbing to the effects of the volcanic ash.

(Below) Birds were some of the first casualties of the ash storm and may have died within hours after the ash began to fall.

Rhino foot bones (right) from the volcanic ashbed. Note the whitish, bumpy bone growth on the toe bones. This is the abnormal, pathologic bone growth that appears to be associated with extensive lung damage from hypertrophic osteopathy.

Photos by Angie Fox, University of Nebraska State Museum

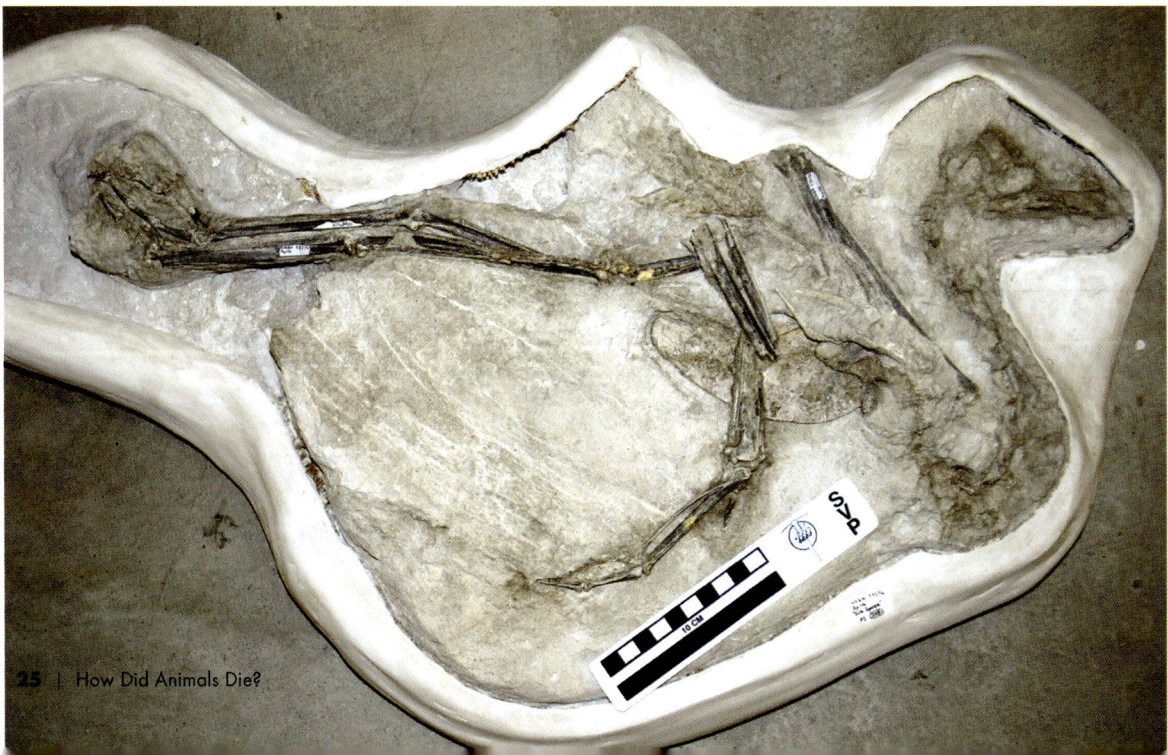

Smooth, healthy bone

Abnormal, bone growth
indicative of Marie's Disease

2 inches

Aspects of Paleoecology

An understanding of the ecologic setting in the Ashfall waterhole is realized by evidence of numerous species of diatoms that are concentrated in the basal two inches of the ashbed, and the uppermost inch of the waterhole sand. Student and faculty researchers from Iowa Lakeside Labs have been sampling for diatoms on an annual basis (Serissoryl, et al., 2003). Along with aquatic vertebrates such as painted turtles (*Chrysemys*), musk turtles (*Sternorthurus*), salamanders (*Ambystoma*), frogs (*Rana*), and the absence of fish (with the exception of a few very small vertebrae that may have been transported to the site by scavengers), the inference is strongly suggestive of an ephemeral lake that dried, on a cyclic, probably seasonal basis. The caliche beds mentioned earlier form in a seasonally arid environment. The habitat nearby appears to have been grassland studded with trees, or intermixed with woodlands. Fossil anthoecia (chaff or hulls) from three species of grass seed and three species of sedge, some of which are found in the mouths or teeth of the rhino and horse skeletons, are commonplace in the ashbed and adjacent strata.

The majority of herbivore skeletons (*Teleoceras*, and all of the equids) are from species with hypsodont molariform dentition, or high-crowned cheek teeth, inferring a grazing or mixed-feeding diet. A few herbivore species such as the three-horned "deer" *Cranioceras* are brachydont with low-crowned cheek teeth, and fed on less abrasive food from forbs and woodland plants. Seeds from two deciduous tree species including the hackberry (*Celtis occidentalis*), and the walnut (*Juglandicarya* sp.) are known from the site.

Plant fossils have been examined by Voorhies and Thomassen since 1979. An overall assessment of the composition of the species and abundance represented by the biota, infer a subtropical savanna with a seasonally arid climate.

Clues about ancient temperature

Looking at the abundant fossils of rhinos, camels and elephants in the Ashfall Fossil Beds, it's hard not to speculate that Nebraska must have been a warmer place to live twelve million years ago. However, maybe one shouldn't overemphasize the fact that the closest living relatives of these animals tend to inhabit the warmer parts of the Earth today. They are mammals, after all, and therefore were "warm-blooded" (able to maintain a high, constant body temperature in all kinds of weather). We should also remember that one of the two living species of Old World camel, the double-humped Bactrian, runs wild in Mongolia which has Wyoming-like winters.

What was the Climate Like?

• It was much milder than present day Nebraska.

• Winters were warmer but summers may have been cooler.

• Total rainfall may not have been much different from today but periodic droughts were a problem then as now.

• The annual cycle may have been dominated by rainy and dry seasons rather than today's cold winters alternating with hot summers.

• There definitely was enough difference between the seasons to affect the growth and reproduction of plants and animals.

If fossil mammals may be unreliable guides for tracking ancient climates, are there other, more reliable clues? Yes. It turns out that giant land turtles (tortoises) are among the most sensitive "paleothermometers" in the fossil record and their remains are common at Ashfall. These tank-like animals survive in the wild today only on tropical islands in the Pacific and Indian Oceans but, more importantly, their "cold-blooded" metabolism can't keep them alive for more than a few days if the air temperature drops to near freezing. Unlike smaller tortoises, like the little box turtles that go into underground burrows to survive a brutal winter in Ashfall Park today, giant tortoises are far too big to burrow. Therefore when paleontologists find giant tortoise fossils, they feel justified in concluding that the climate was frost-free when and where the animals lived.

Close inspection of the shells of Ashfall's giant tortoises yields another important insight into the ancient climate: that it was strongly seasonal. Instead of being smooth, as they would be if the tortoises grew at the same rate throughout the year, the separate bony shell plates have concentric growth rings. By analogy with such rings in modern turtles, these are interpreted as evidence of alternating periods of fast and slow growth. Whether these growth spurts were tracking annual fluctuations in temperature or perhaps in rainfall is not known.

Bottom line - Precipitation

If tortoise shells are like ancient thermometers buried in the Ashfall time capsule, how about an ancient rain gauge? Are there any reliable clues about how much rain the Ashfall animals may have experienced in their lifetimes? Beyond the fact that there must have been enough rain to support plants to feed a huge abundance and diversity of plant-eating animals, the animal fossils don't directly tell very much. The fossil plants at Ashfall, however, are very informative. Paleobotanist Joe Thomasson has identified ten species of fossil plants from the sandstone below the ash (see list in appendix). The great majority of the species are grasses, but fossil seeds of the hackberry tree and one petrified walnut shows that some trees were present also. A grassland with some trees is referred to as a savanna. Probably no more than about thirty inches of rain fell per year. More than that, and trees would begin to dominate.

In recent times, about twenty percent of Nebraska's moisture comes in the form of snow and the Ashfall animals surely did not experience that since air temperatures didn't drop below freezing. In present day Nebraska most of the rain comes during spring and early summer; late summer and fall are usually much drier. Reproduction in many present-day plains animals is strongly tied to this cycle: bison, deer, elk, and pronghorns give birth in the spring when the grass greens up and is available for new mothers and their offspring.

How Long Did it Take Before Life Returned to the Devastated Area?

Fossils discovered above the ashbed prove that the savanna eventually came back to life, but how fast and how complete the recovery was are still questions under active study.

No skeletons have been discovered in the upper part of the volcanic ashbed so we can conclude that no large animals were left alive in the area, except perhaps for a tortoise or two, after the last rhino died. At the very top of the ashbed are some fossil tracks, some elephant-sized some smaller, an indication that a few animals walked across the area where the waterhole used to be.

Mural by Mark Marcuson

Paleontology of the Ashbed

The Ashfall site is widely recognized in the scientific community as an exceptional natural occurrence. It is referenced by authorities in the field of paleontology, such as Bruce MacFadden of the University of Florida who describes the Ashfall Site as "*Lagerstatten* (or mother lode), a locality featuring both extraordinary preservation and abundance . . . The scientific value of these localities as unique opportunities to better understand the paleobiology of past life is extraordinary. Gould, 1989, says that '*Lagerstatten* are rare, but their contribution to our knowledge of life's past history is disproportionate to their frequency by orders of magnitude.'. . . we shall examine six truly extraordinary fossil horse localities . . . of these perhaps only Messel (Germany) and the Ashfall Fossil Beds in Nebraska qualify as *Lagerstatten*." (MacFadden, 1992)

The faunal remains reflect an assemblage of the Clarendonian Land Mammal Age (Tedford, et al., 1987, Voorhies, 1985, 1990b). This "age" refers to the species from ten to twelve million years before present. A species by species description of the Ashfall animals is included in the following pages of this publication. Here is a brief overview of the more commonly found animals:

Rhinos

The most numerous remains in the site are from the short-legged and barrel-bodied one-horned rhinoceros *Teleoceras major*, the males of which stood only about forty-two inches tall at the shoulder while having a ten-foot chest circumference. More than one hundred individuals have been uncovered in the ash of the ancient waterhole and many of these are intact articulated skeletons preserved three-dimensionally with full, rounded shape of the rib cages and skulls. *Teleoceras* may have frequented bodies of water like the extant hippopotamus (*Hippopotamus amphibius*). The presence of silicified grass anthoecia (seed chaff) in their mouths and abdomens confirm a grazing diet (Voorhies and Thomasson, 1979).

Barrel-bodied Rhino
Teleoceras major

Teleoceras major is, by far, the most abundant mammal within the ash horizon, and it is likely that most, if not all, of the members of a herd were preserved at the site. This accumulation of associated articulated skeletons allows an assessment of gender and sexual dimorphism as well as the age and social structures of the herd.

Gender is easily ascertained from the dental characteristics of *Teleoceras major*. H. F. Osborn's conclusion that skulls of *Teleoceras* could be assigned a gender on the basis of the size of the lower second incisor (Osborn, 1898) was verified at Ashfall when fetal bones were found in the pelvic region of an individual with small incisors (Voorhies and Stover, 1978). Thus, Voorhies (1985) was able to identify thirty-two adult females and only five adult males, noting underrepresentation of young adult males. Furthermore, individuals with smaller incisors were proportionately smaller than individuals with large tusks, strongly suggesting sexual dimorphism. Alfred J. Mead (2000) verified this sexual dimorphism with a suite of cranial and limb measurements, finding that the most dimorphic features were lower incisor diameter (the second lower incisor is the tusk in *Teleoceras*), and appendicular elements such as the radius mid-shaft cross-sectional area, femoral cross-sectional area beneath the third trochanter, and thickness of the third metacarpal. The morphological variability within the appendicular skeleton reflects adaptations toward a greater body mass in males. Mead (2000) estimated the body mass for male *Teleoceras major* at Ashfall to be roughly 1,940-2,450 pounds and for females 1,730-1,850 pounds.

The age structure of the Ashfall *Teleoceras major* herd can be estimated using tooth eruption and wear criteria that were originally developed for assessing populations of the extant black rhinoceros, *Diceros bicornis* (Goddard, 1970; Hitchins, 1978). The application of these criteria to the Ashfall herd identifies a minimum of ten distinct age classes ranging from young calves to old adults (Voorhies, 1985; Mead, 1999). The youngest individuals fell into three distinct age classes with little or no variation in the degree of tooth eruption or occlusal wear suggesting that birthing was seasonal in *Teleoceras major*, unlike the extant African rhinos (Voorhies, 1985). The oldest individuals preserved in the death assemblage are dentally comparable to those in Goddard's (1970) age class that is indicative of individuals between thirty to forty years old. Voorhies (1985) interpreted the grossly disproportionate ratio of twenty-five young female to one young adult male as evidence that subordinate males were excluded from the herd.

A second rhinocerotid taxon, the long-legged and hornless *Aphelops*, has been recovered from the sands beneath the ashbed. *Aphelops* is rare in most late Miocene deposits on the Great Plains.

Hornless Rhino
Aphelops

Horses

The five genera of horses represent two tribes within the subfamily Equinae. *Cormohipparion*, *Neohipparion*, and *Pseudhipparion* are within the Tribe Hipparionini (MacFadden, 1992), the species of which are characterized by three-toed (tridactyl) feet, which are maintained until the last of the hipparions went extinct in North America in the late Pliocene, three million years ago, and in Asia during the Pleistocene, approximately one million years ago.

Pliohippus and *Protohippus* are representatives of the Tribe Equini which gave rise to *Equus* and the modern species of horses (McFadden, 1992). Twelve million years ago, *Pliohippus* was in a transitional phase between tridactyly and monodactyly. Ashfall skeletons of *Pliohippus* show varying degrees of "three-toedness", some specimens exhibiting side-toes that are reduced to a single-toed (monodactyl) stance.

Sturdy Three-toed Horse
Cormohipparion

Slender Three-toed Horse
Neohipparion

Small Three-toed Horse
Pseudhipparion

Stout One-toed Horse
Pliohippus

Slender, Grass-clipping Horse
Protohippus

Photo by the Nebraska Game and Parks Commission

Front feet of five different Miocene horses from the Ashfall site contrast with the same bones of a modern horse (far right). The first three are typical three-toed horses with well-developed side hooves extending down on each side of the main hoof. Three-toed horses were probably good at cornering and might have spent considerable time in woody or marshy terrain. The next two are functionally single-toed, although only the fifth has lost all traces of the side hooves. Shriveled, vestigial side hooves are still present on the fourth. Like modern single-toed horses, ancient horses probably excelled at flat-out running on the open grassland. From left, *Pseudhipparion*, *Cormohipparion*, *Neohipparion*, *Protohippus* and *Pliohippus*. The genus *Equus* includes all recent horses, asses and zebras.

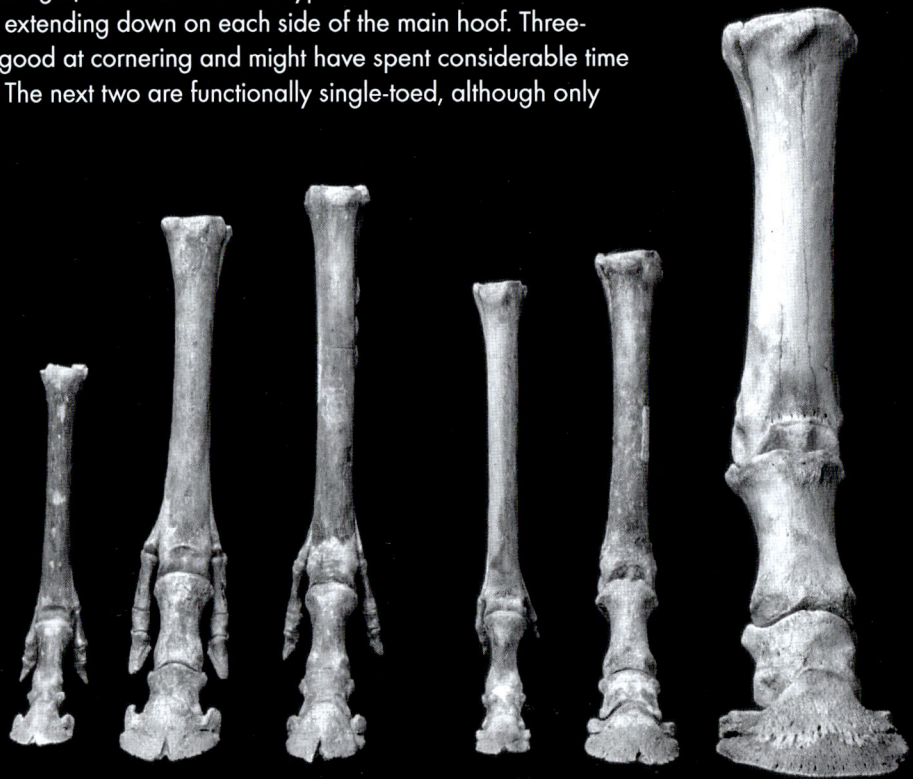

Camels

One of the camelid genera, *Procamelus*, is considered ancestral to the extant Old World camels. A camel genera common on the Great Plains at Ashfall time, but not found as of yet at the site, *Hemiauchenia*, is ancestral in lineage to the South American llamas and their wild kin—guanacos and vicunas.

Ancestral Camel
Procamelus

Primitive Deer

The musk deer *Longirostromeryx* is a primitive antlerless deer (family Moschidae that still inhabit east Asia). Male musk deer are characterized by elongated upper canine teeth, hence the nickname "saber-tooth deer."

Saber-tooth Deer
Longirostromeryx

Carnivores

Three of the six "dog" species of the family Borophaginae, are large bone-crushers capable of scavenging large ungulate carcasses. Evidence of *Epicyon haydeni*, *Epicyon saevus*, and *Aelurodon* (species not determined), at the Ashfall site are limited to isolated bones from the waterhole sand, scavenged bone in the ashbed, and coprolites (fossilized droppings) that contain small shards of bone. Elliptical disturbances filled with a sand/ash mix, in the lower three feet of the ashbed ranging in size from 10 inches diameter to 16 inches by 30 inches in diameter closely resemble fossil burrows, probably dug by the larger borophagines (or the Amphicyonid mentioned below) in search of buried carcasses. A complete skeleton of the small omnivorous *Cynarctus* was exposed in the ashbed within the Hubbard Rhino Barn in 2010. This represents the only intact skeleton of a carnivore found in the ash as of date of publication. A few isolated bones of another small dog, *Leptocyon* are also known from the ash. The coyote-sized *Carpocyon* (possibly a fruit-eating species) is known from a toothless lower jaw recovered from the waterhole sand.

Interestingly, another large carnivore with bone-crushing dental adaptations inhabited the area and is the beardog *Ischyrocyon gidleyi*, an amphicyonid that weighed in the four to six hundred pound range. *Ischyrocyon gidleyi* is only known from isolated elements recovered from the waterhole sand.

Giant Bone-crushing Dog
Epicyon

Bone-crushing Dog
Aelurodon

Raccoon-like Dog
Cynarctus

Beardog
Ischyrocyon

Fruit Dog
Carpocyon

Dog Ancestor
Leptocyon

Snake

An exciting and unusual find in 2008 was an articulated snake, complete from the skull bones to about mid-section of the torso. The specimen, a common constrictor (family Colubridae) is currently under study.

Birds

Three described avian species from the site are new to science. Relatively rare in the fossil record, the gentle, rapid burial preserved the delicate bones, even tracheal cartilage, ossified tendons, and a feather impression or two. The three species referred to are: crowned crane (*Balearica exigua*), long-legged hawk (*Apatosagittarius terrenus*), and Old World vulture (*Anchigyps voorhiesi*)(Feduccia, 1996, Feduccia and Voorhies, 1989, Zhang, et al., 2012).

Eagle-like scavenger
Anchigyps voorhiesi

Crowned Crane
Balearica exigua

Secretarybird Mimic
Apatosagittarius terrenus

Portraits

All of the different species of fossils (including plants as well as animals) found in the Ashfall Fossil Beds through the 2012 excavation season are listed on the last page of this booklet. Probably more interesting to most readers are the portraits of thirty-six species of extinct animals found at Ashfall shown on the following pages. They were painted by artist Mark Marcuson between 2009 and 2010, and are based on careful direct study of fossil remains along with knowledge of the anatomy and habits of living animals.

Since only hard parts (bones, teeth, and sometimes cartilage) are preserved in available fossils, the colors and skin patterns of the reconstructed animals are based on educated guesses. For instance, the Ashfall rhinos and elephants are painted gray because all species of modern day elephants and rhinos are gray. The extinct barrel-bodied rhino is shown with a horn even though the horn itself (made of hardened skin) is not preserved in Ashfall skeletons because it presumably rotted away along with other soft edible tissues during fossilization. We know for sure this rhino had a horn, however, because the nasal bones of the species have a roughened area where the horn was attached. Likewise, black shiny hooves are shown on the extinct horses even though only the bones inside the hooves are preserved in the Ashfall horse skeletons. Such details are based on the appearance of the closest living relatives of the fossil species.

Note that not all of the illustrated species are represented by whole skeletons found in the volcanic ashbed. Some animals are known (so far) only from fossils found in the sandstone directly beneath the volcanic ashbed. Reconstructions of these species, like the beardog and the ancestral kangaroo rat, are sometimes based on more complete fossils found at other sites.

Our portrait gallery is not complete. A number of small animals (especially rodents) are not yet well enough known to be illustrated. New discoveries, including more complete remains of species already identified, continue to be made as more digging occurs in the park. Future editions of the booklet will have pictures of additional Ashfall species.

Barrel-bodied Rhino
Teleoceras

Adult males: 3½ feet (1.1 meters) tall at shoulder, 10 feet (3 meters) long, 10 feet around the belly.
Females: about 20% smaller.

The most abundant large animal discovered in the volcanic ashbed is an extinct rhinoceros with a body shape similar to today's hippopotamus. Like hippos, barrel-bodied rhinos may have wallowed in the waterholes that dotted the ancient savannas of the Great Plains. Unlike modern-day rhinos, which are primarily solitary creatures, the Nebraska rhino probably was a social species that formed herds. By studying the age and sex of more than one hundred skeletons from Ashfall, paleontologists concluded that *Teleoceras* males (with large tusks) may have defended "harems" of females (with small tusks) and their calves. Young adult male skeletons are remarkably rare in the ashbed, suggesting that "bachelor males" may have been excluded from the breeding herds and forced to live elsewhere.

Pronounced **Tee**-lee-**oss**-ur-us, meaning "perfect horn" and refers to the fact that both male and female skulls have rough bumps near the end of their nasal bones proving that they had a true horn in life.

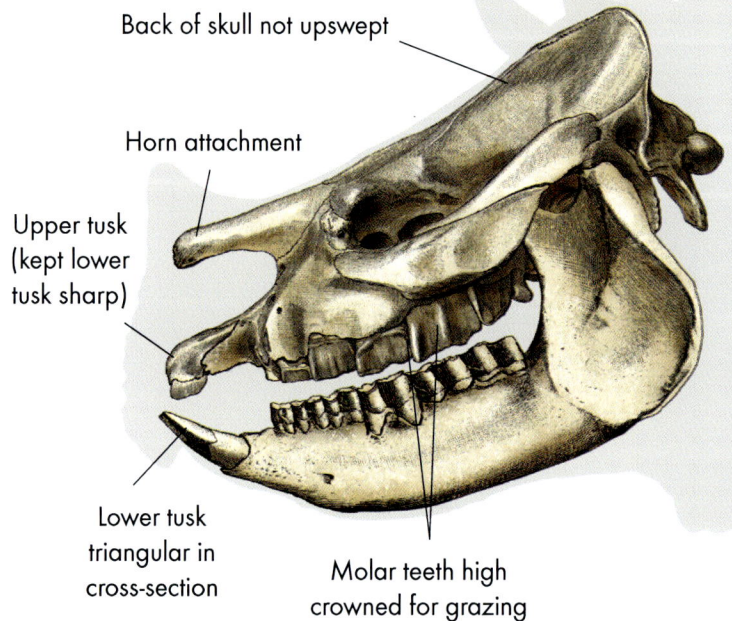

Back of skull not upswept

Horn attachment

Upper tusk (kept lower tusk sharp)

Lower tusk triangular in cross-section

Molar teeth high crowned for grazing

Remains of this animal have been found in the:
"RECOVERY" layer (sandstone above ash)
"DISASTER" layer (volcanic ashbed)
"WATERHOLE" layer (sandstone below ash)

Height: The Ashfall species of *Aphelops* stood about 3 1/2 feet (1.15 meters) at the shoulder.

Two very different species of rhinos lived in the Great Plains between twelve million and five million years ago. One was the hippo-like grazer *Teleoceras*—the barrel-bodied rhino so common at Ashfall—and the other was the more normally proportioned *Aphelops* which lacked a horn.

Judging from the sparse record here in Ashfall Park (a few teeth and bones from the "waterhole" sand and a lower jaw from the "recovery" layer) hornless rhinos rarely visited the Ashfall waterhole. *Aphelops'* teeth suggest it browsed on woody vegetation like African black rhinos do today. Since the habitat around the ancient waterhole was mostly grassland, hornless rhinos and other browsers probably found the pickings pretty slim around here twelve million years ago.

Rhino horn is made of tough but edible skin so it is rarely fossilized, but skull bones beneath the horn in modern rhinos have a distinctive texture. Therefore paleontologists can tell whether an extinct species had a horn or not.

Pronounced **Af**-ful-ops, the name means "smooth-faced" or "hornless."

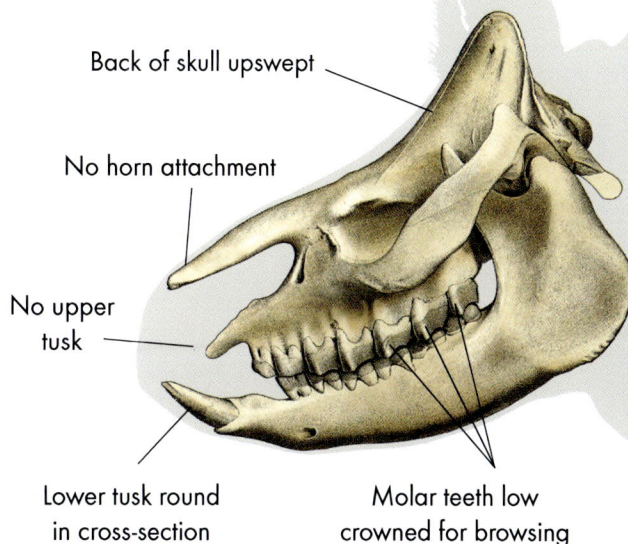

Back of skull upswept

No horn attachment

No upper tusk

Lower tusk round in cross-section

Molar teeth low crowned for browsing

Remains of this animal have been found in the:
"RECOVERY" layer (sandstone above ash)
"WATERHOLE" layer (sandstone below ash)

Slender Three-toed Horse
Neohipparion

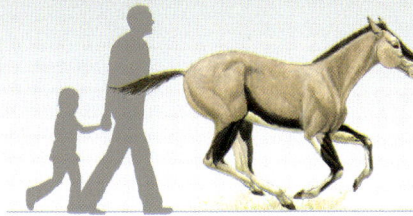

Height: 4 feet (1.2 meters) at the shoulder

The long, slender legs of this three-toed horse show that it was a fast runner. The large functional "side hooves" (one on each side of the main hoof) suggest that it was a good dodger as well and perhaps was better at dealing with soft, treacherous ground than the single-toed horses. Like the other four species of horses found in the ashbed, this one has tall-crowned molar teeth capable of grinding tough, silica-rich grass into digestible mush. The species found at Ashfall, *Neohipparion affine*, is larger than its ancestors, found in older fossil beds, and smaller than its descendants, found in younger deposits.

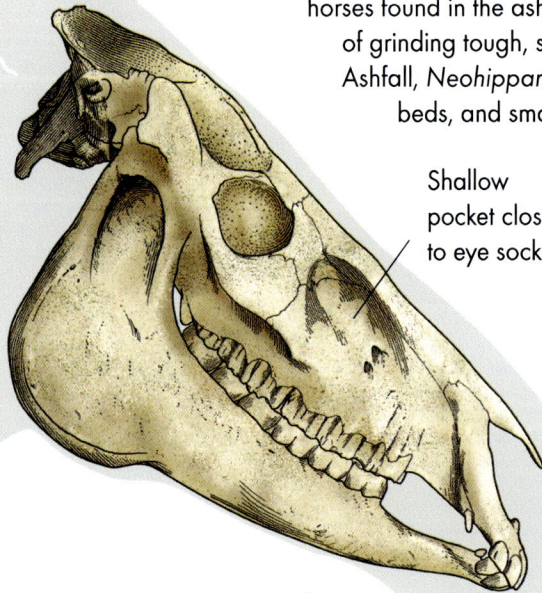

Shallow pocket close to eye socket.

When the first fossils of three-toed horses were found in Europe they were named *Hipparion*. When a skeleton of a similar horse was found later in South Dakota it was named *Neohipparion* ('neo' means new) because of its occurrence in the New World. The greek word for horse is "hippos" and "-arion" means "small" or "diminutive."

Single teeth of *Neohipparion* can be difficult to distinguish from those of *Cormohipparion*, hence the question marks about its occurrence above and below the ashbed.

Remains have been found in the:
"RECOVERY" layer (sandstone above ash) ?
"DISASTER" layer (volcanic ashbed)
"WATERHOLE" layer (sandstone below ash) ?

Height: about 3½ feet (1.1 meters) at the shoulder.

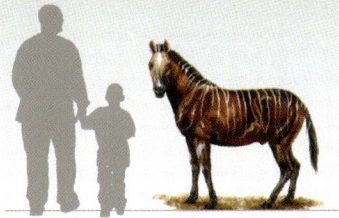

Sturdy Three-toed Horse
Cormohipparion

Nebraska-born paleontologist Morris F. Skinner collected and studied fossil horses for more than fifty years. One of his greatest discoveries was a series of well-preserved skulls of Nebraska horses apparently ancestral to the "*Hipparion*" horses of Europe and Asia. Other paleontologists had been forced to rely mostly on teeth in researching horse evolution but Skinner pointed out important details of the skull—especially the position and shape of a depression in front of the eye socket. Announcing this discovery in 1977, Skinner and a younger colleague Bruce MacFadden named *Cormohipparion*. The first complete skeletons of this animal have been found here at Ashfall. These show that the foot bones of *Cormohipparion* were shorter than those of *Neohipparion*, a horse with very similar teeth.

The Ashfall species, *Cormohipparion occidentale*, has one of the most complicated enamel patterns on the grinding teeth of any North American fossil horse species. This would help grind grass into small, digestible fragments. *Cormo* means "stem" and *Hipparion* is the name of the three-toed grazing horse of the Old World.

Chewing surface of upper molar tooth.

A deep pocket in the skull far forward of the eye is characteristic of this horse.

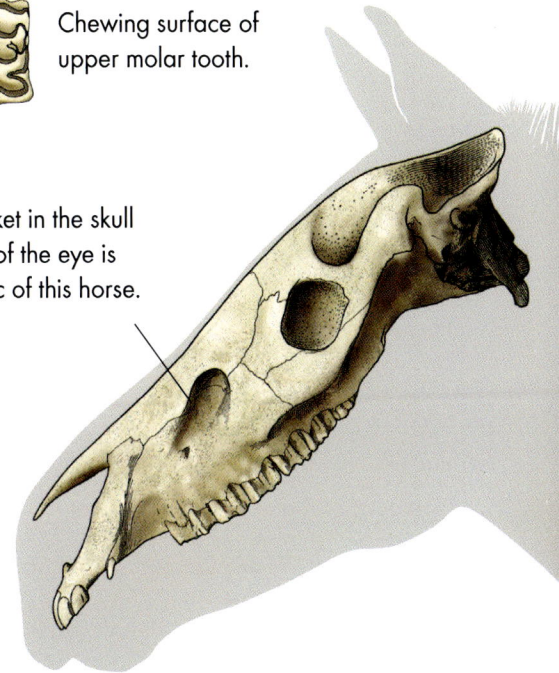

Remains of this animal have been found in the:
"RECOVERY" layer (sandstone above ash)
"DISASTER" layer (volcanic ashbed)
"WATERHOLE" layer (sandstone below ash)

Small Three-toed Horse
Pseudhipparion

Height: 2 1/2 feet (0.75 meters) tall at the shoulder.

A shallow depression in this position identifies the skull of this small horse.

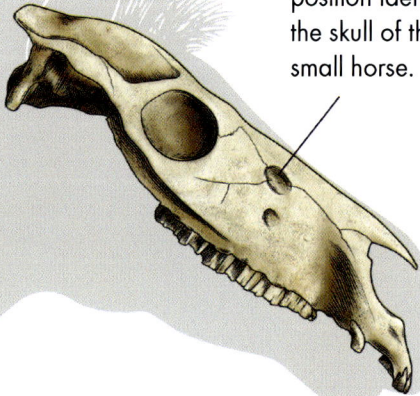

Not much bigger than a large dog, *Pseudhipparion* was the smallest of the Ashfall horses. It was also the most abundant: remains of more than fifty individuals, from newborns to old adults, have been excavated here. Well-developed side-hooves probably helped the animal gain traction on soft ground and in making quick turns. The teeth are typical of a grazer (grass-eater) but have peculiarities which prove that this little horse is not closely related to other three-toed horses but was on its own evolutionary path. *Pseudhipparion* originated about fourteen million years ago and became extinct ten million years ago, generally becoming smaller in body size throughout this time range, differing from most horse lineages in this respect. The Ashfall species *P. gratum* is of average size.

Pronounced **Sood**-hip-**pair**-e-on, the name means "misleadingly similar to *Hipparion*" the Old World three-toed horse.

Remains of this animal have been found in the:
"RECOVERY" layer (sandstone above ash)
"DISASTER" layer (volcanic ashbed)
"WATERHOLE" layer (sandstone below ash)

Height: 3½ feet (1.1 meters) at the shoulder.

Stout One-toed Horse
Pliohippus

The largest single-toed horse found at Ashfall, *Pliohippus* was about the size of a big white-tail or mule deer. Like the other four kinds of horses excavated from the ashbed, Pliohippus was a grazer that probably lived in social herds as modern horses do. When the first partial skeleton of *Pliohippus* was discovered near Valentine, Nebraska, in 1873, it created excitement in the world of paleontology because it was the last major "missing link" in the evolution of the horse to be discovered. By then, the tiny horses from thirty to fifty million years ago were already known and sheep-sized three-toed horses had been collected from fifteen to twenty million-year-old beds but no fossils were known that demonstrated how the vestigial "extra" toes were eliminated.

Hippos means horse and *Plio* refers to the Pliocene Epoch of geologic time, two to five million years ago.

A pair of deep, pocket-like depressions in the skull ahead of the eye socket distinguishes *Pliohippus* from other horses.

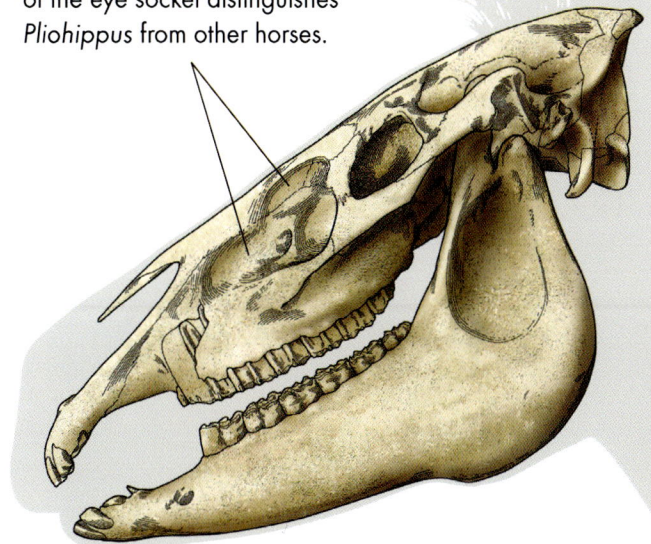

Remains of this animal have been found in the:
"RECOVERY" layer (sandstone above ash)
"DISASTER" layer (volcanic ashbed)
"WATERHOLE" layer (sandstone below ash)

Slender, Grass-clipping Horse
Protohippus

Height: about 3 feet (0.9 meters) at the shoulder. Its legs were proportionately as long as the living wild ass of Somalia, a very fleet animal.

Arrangement of the Incisor (Nipping) Teeth

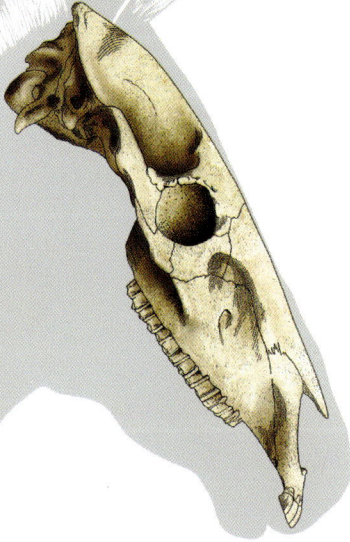

Curved
In *Pliohippus* and most other fossil and recent horses

Straight
In *Protohippus* (and *Calippus*)

Despite its name, *Protohippus* was not the first (oldest) member of the horse family but it does share some skeletal and dental features with modern horses that other twelve-million-year-old kinds of horses do not. For example, it has a "smooth face" (i.e. no deep depressions in the skull in front of the eye sockets). Its chewing teeth are also more similar to modern horse teeth than are those of the other four horses found in the ashbed. Its front teeth, however, are more specialized than those of living horses. They are arranged in a straight line presumably to clip more grass at ground level than other horses which have the nipping teeth arranged in a curve. Like *Pliohippus*, *Protohippus* had vestigial, nonfunctional side toes and probably was a fast runner in flat open grasslands.

Protos means "the very first" and *hippos* means "horse."

Protohippus could clip grass close to the ground.

Remains of this animal have been found in the:
"DISASTER" layer (volcanic ashbed)
"WATERHOLE" layer (sandstone below ash)

Height: These heavy-bodied horses stood about 3 1/2 feet (1.1 meters) tall at the shoulder.

A single molar tooth, excavated in 1978, is the only evidence that browsing horses used the Ashfall waterhole. Twelve million years ago *Hypohippus* could have qualified as a "living fossil"—the last surviving member of an evolutionary line extending back more than twenty-five million years. Except for larger size, the bones and teeth of *Hypohippus* closely resemble those of its ancestor *Mesohippus*, known from the Badlands of South Dakota and Nebraska. These early horses had short-crowned teeth for eating leaves and short legs with three well-developed hooves on each foot. They probably lived in forested areas with plenty of browse. As the climate dried and grasslands expanded, browsing horses became scarcer and eventually died out between ten and eleven million years ago.

The eye socket is farther forward than in grazing horses (situated directly above the wisdom tooth rather than behind it). The muzzle is very short compared to other horses.

Chewing teeth with strong roots and short enamel crowns are suitable for eating leaves not grass.

Hypo means "below" and *hippos* is Greek for "horse."

Remains of this animal have been found in the:
"WATERHOLE" layer (sandstone below ash)

Oreodont
Ustatochoerus

Height: about 2 feet (0.6 meters) at the shoulder.

Oreodonts have been extinct for ten million years and have no close living relatives so they remain mysterious in many ways. Between thirty-five million and twenty million years ago, they were the most abundant mid-sized mammals over much of North America. Astounding numbers of skulls have been collected from the northern Great Plains. This great abundance misled early paleontologists into thinking that oreodonts formed vast herds like bison. Recent research, however, shows that they lived mostly in small family groups and were not grazers. Oreodonts probably used their short, flexible limbs for climbing trees and their sharp, interlocking front teeth for cutting branches. When forests began to disappear from the Great Plains about fifteen million years ago, oreodonts became scarce.

Oreodont means "mountain tooth" and refers to the sharp-crested appearance of the chewing teeth as seen from the side.

Remains of this animal have been found in the:
"RECOVERY" layer (sandstone above ash)
"WATERHOLE" layer (sandstone below ash)

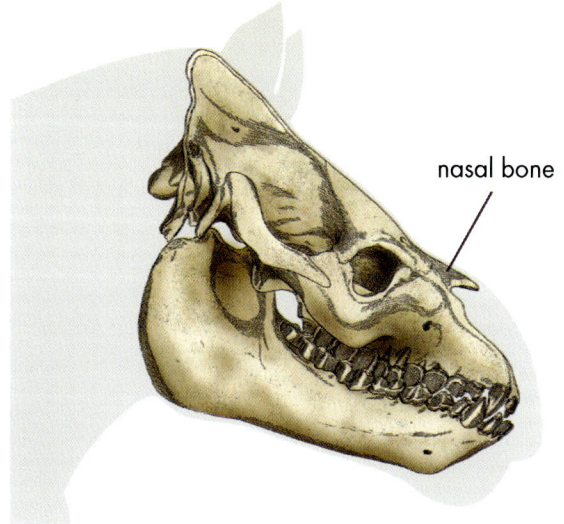

nasal bone

A posterior position of the nasal bone is associated with a flexible snout in modern animals. Razor-sharp interlocking front teeth worked as excellent branch-cutters. There is no gap between the front teeth and chewing teeth (more similar to primates than to modern grazing animals like bison or horses).

Height: about 11 feet (3.3 meters) tall at the head.

Giraffe Camel
Aepycamelus

An Example of Convergent Evolution
Unrelated animals develop similar anatomy in response to similar lifestyle.

Giraffe-camel
(neck and hind leg)

Modern Giraffe

Only **four** of the neck vertebrae are elongated.

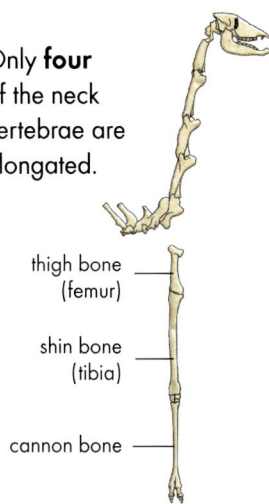

Six of the seven neck vertebrae are elongated about equally.

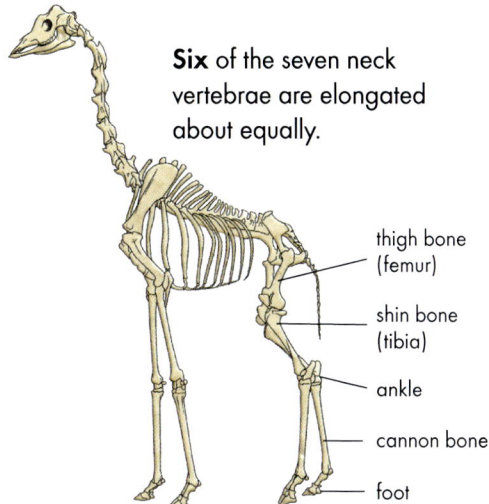

thigh bone (femur)

shin bone (tibia)

cannon bone

thigh bone (femur)

shin bone (tibia)

ankle

cannon bone

foot

In giraffe-camels, the shin bone is longer than the cannon bone.

In giraffes, the cannon bone is longer than the shin bone.

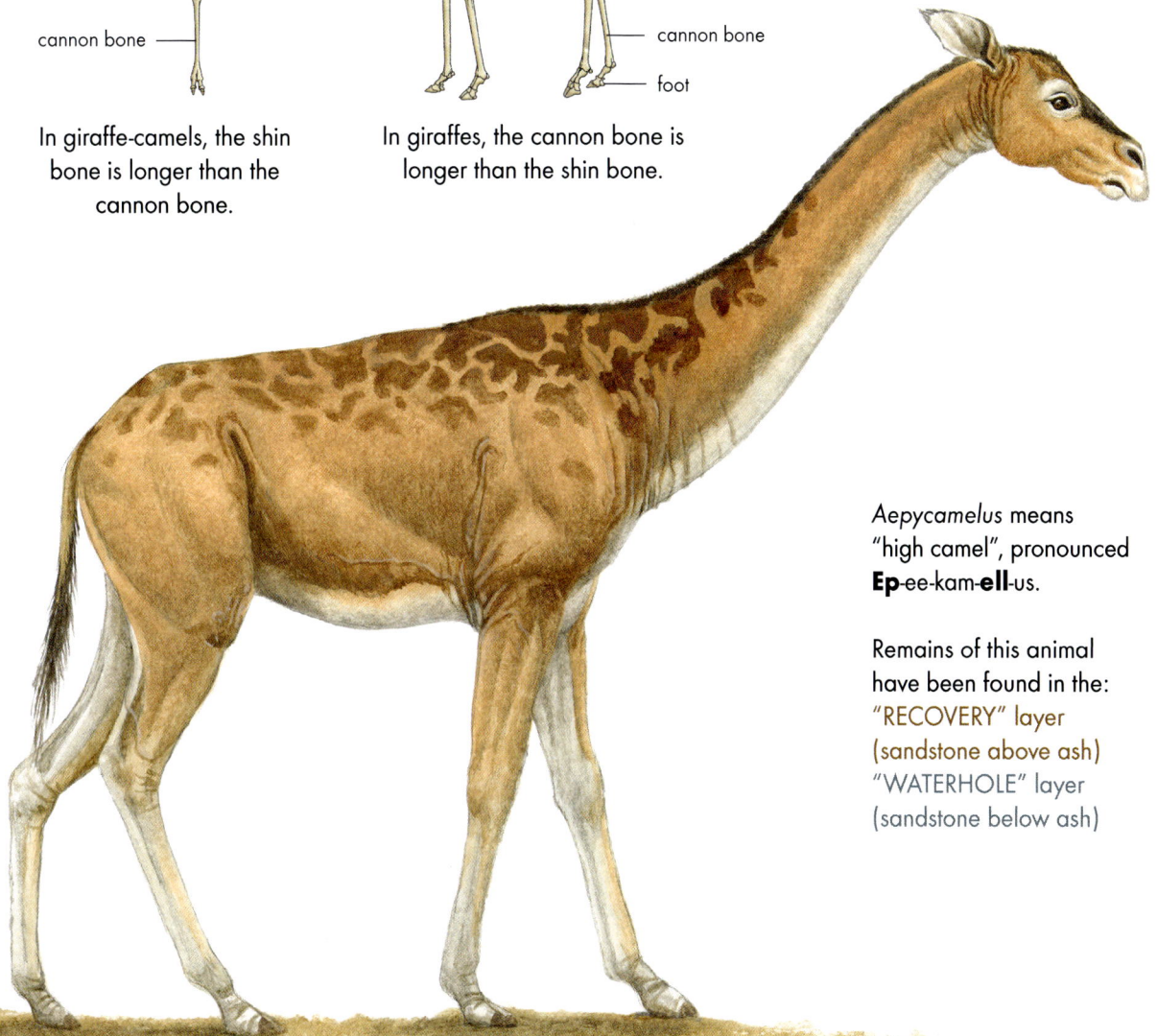

True giraffes never lived in America but one branch of the camel family evolved very long legs and necks and must have looked much like giraffes do today. They probably browsed on tree leaves out of the reach of other hoofed animals. Like giraffes, they probably had a hard time getting their lips in the water to drink.

Aepycamelus means "high camel", pronounced **Ep**-ee-kam-**ell**-us.

Remains of this animal have been found in the:
"RECOVERY" layer (sandstone above ash)
"WATERHOLE" layer (sandstone below ash)

Ancestral Camel
Procamelus

Height: Adult males were about 5 feet (1.5 meters) tall at the shoulders. Females were slightly smaller.

Except for horses, camels were the most abundant medium-sized animals found at the Ashfall Fossil Beds. Remains of several dozen individual camels—from young calves to adults of both sexes—have been found in the volcanic ashbed. Apparently an entire herd of medium-sized camels died in the waterhole, mostly belonging to the species *Procamelus grandis*. Detailed studies of the skeleton and skull of this species support the idea that it is the ancestor of the one-humped and two-humped camels that live in Asia and Africa today.

Procamelus means "before" or "ancestral to" the camel, pronounced **Pro**-kam-**ell**-us.

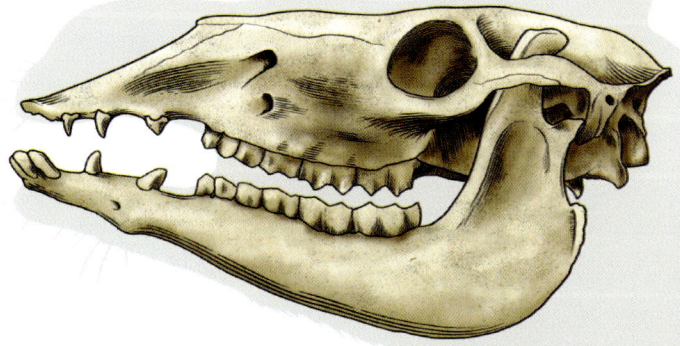

Remains of this animal have been found in the:
"RECOVERY" layer (sandstone above ash)
"DISASTER" layer (volcanic ashbed)
"WATERHOLE" layer (sandstone below ash)

Height: slightly more than 3 feet (1 meter).

Llama-sized Camel
Protolabis

The smallest species of camel so far discovered at Ashfall was about the size and shape of a small llama. Details of the skull and feet, however, show that the resemblance to llamas is superficial and that *Protolabis* is not directly ancestral to llamas. The very narrow muzzle of this camel indicates it had very strong lips and cheek muscles. It probably fed on tough, low-growing vegetation, mostly shrubs.

Protolabis has teeth (shown in white) which were lost during the evolution of modern camels and llamas. Pronounced **Pro**-toh-**lay**-bis, the name means "primitive lips", a reference to the presence of upper incisor teeth in this kind of camel.

Remains of this animal have been found in the:
"RECOVERY" layer (sandstone above ash)
"DISASTER" layer (volcanic ashbed)
"WATERHOLE" layer (sandstone below ash)

Three-horned "Deer"
Cranioceras

Today's deer, elk, and moose grow a new pair of branching antlers every year but the extinct otherwise deer-like creatures found in American pre-Ice Age fossil deposits had very different horns. All three horns were permanently attached to the skull and were probably covered in life by tough, hornlike skin. Males probably used their front horns for combat and to intimidate other males. Females were hornless except for a slight protuberance at the back of the skull. Their bodies were essentially deer-like except for slightly shorter lower legs. Their teeth and jaws are adapted to eating soft, leafy vegetation, not grass. This probably explains why, like other browsers, they were rare in the vicinity of the Ashfall waterhole.

Remains of this animal have been found in the:
"RECOVERY" layer (sandstone above ash)
"WATERHOLE" layer (sandstone below ash)

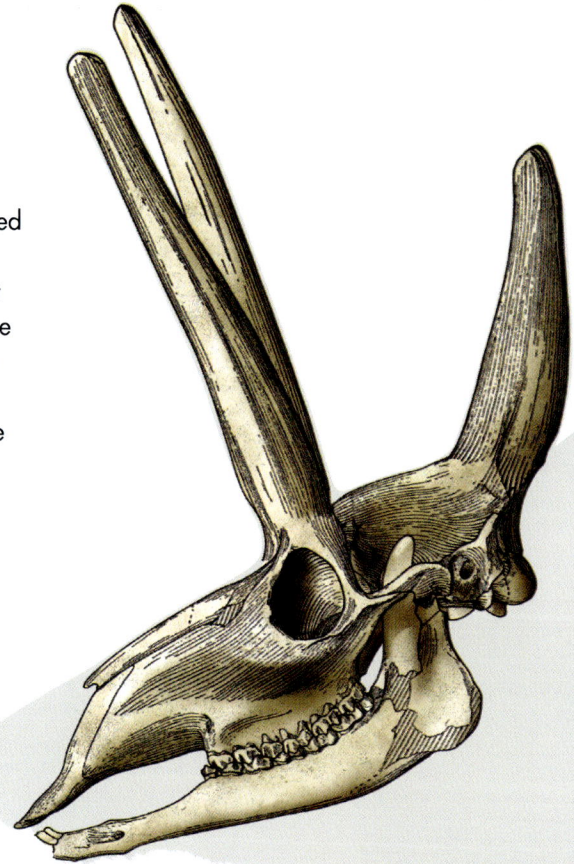

Teeth and jaws are remarkably similar to today's deer but horns are entirely different. Pronounced **Crane**-ee-oh-sir-us, the name refers to the presence of horns (ceras) on the skull (cranium).

Saber-tooth Deer
Longirostromeryx

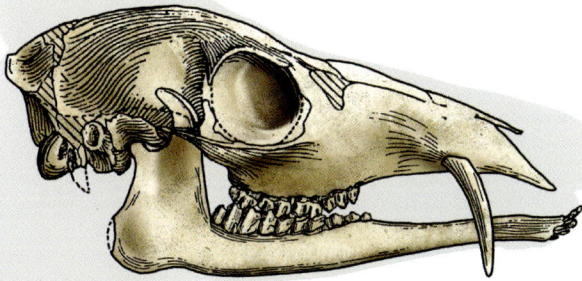

The first thing that comes to mind when we hear the term "saber-tooth" is the famous "saber-tooth tiger" found in Ice Age fossil deposits. But other animals occasionally have long sharp canine teeth too, such as the musk deer that lives in forests and shrublands of southeastern Asia today. Male musk deer are hornless but have saber-like canines that they use for fighting other males in the breeding season. Among the first mammals that died in the ash storm at the Ashfall waterhole were extinct relatives of the musk deer, which had a long, slender muzzle. Modern deer and antelope that have similar slender snouts and delicate incisor teeth are selective browsers on soft vegetation. Tender leaves and shoots were probably the preferred diet of these little browsers.

A long slender snout allowed the animal to poke its nose into narrow places to select the most nutritious tender plants. Short-crowned molar teeth are typical of browsers that eat soft, leafy vegetation. Only males had "fangs".

Pronounced **Lon**-gee-**ross**-trow-**mare**-ix, the name means "long-snouted cud chewer."

Remains of this animal have been found in the:
"RECOVERY" layer (sandstone above ash)
"DISASTER" layer (volcanic ashbed)
"WATERHOLE" layer (sandstone below ash)

Ancestral Prongbuck
Proantilocapra

Prongbucks, sometimes called "antelopes" or "pronghorns," are the fastest species of land animal in the Western Hemisphere; only the cheetah has been clocked at higher top speeds. Unlike other native American grazing animals, like horses and camels, prongbucks never migrated to other continents but remained closely tied to one habitat: the semiarid Great Plains and Great Basin of western North America. During their more than twenty million years on American soil, the prongbuck family diversified into dozens of branches only one of which survives today.

Pro means "before" and *Antilocapra* (literally "antelope-goat") is the scientific name of today's prongbuck.

Unlike deer, which shed their bony antlers every year, prongbucks have a permanent bony horn-core perched above the eye socket. The horn itself is like a hollow "glove" which falls off and has to be regrown annually.

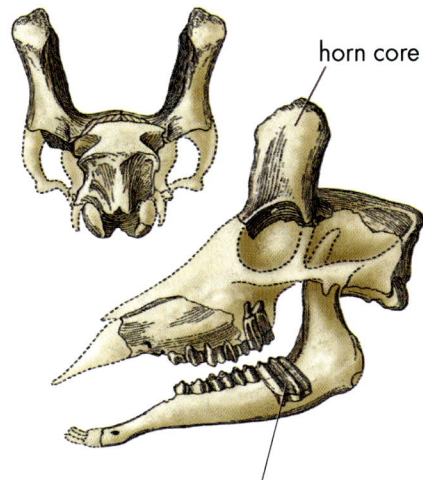

horn core

Prongbuck teeth are very tall-crowned and can deal with very tough grass and sage brush.

Remains of this animal have been found in the:
"RECOVERY" layer (sandstone above ash)
"WATERHOLE" layer (sandstone below ash)

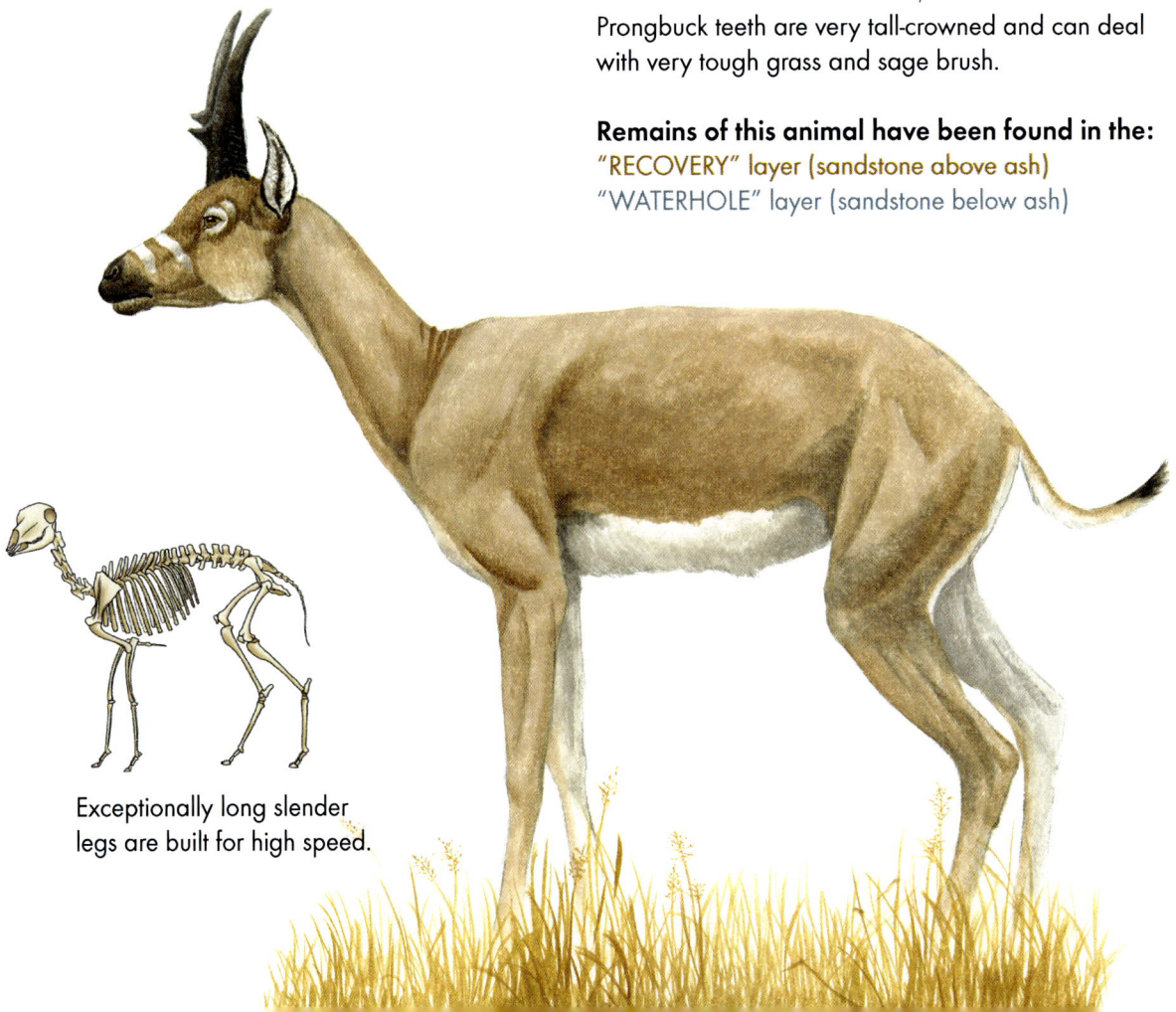

Exceptionally long slender legs are built for high speed.

Height: The Ashfall species, *Aelurodon taxoides*, stood about 2 1/2 feet (0.7 meters) at the shoulder.

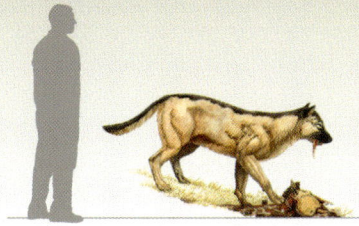

Bone-crushing Dog
Aelurodon

tall crest for attachment of jaw muscles

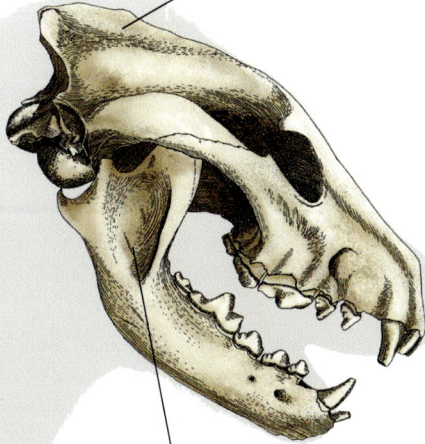

deep pocket for insertion of jaw-closing muscles

The largest members of the dog family alive twelve million years ago are classified in a totally extinct subfamily called the bone-crushers that were not ancestral to modern wolves, coyotes, and dogs. Compared to a wolf of similar size *Aelurodon* had more massive jaws and teeth and slightly shorter feet. The shape of the bones in the foreleg indicate that the animals could rotate their front feet more flexibly than modern dogs can. Their top speed was probably less than a modern wolf but they are likely to have been active predators able to chase down medium-sized horses and camels. Many single bones in the Ashfall Fossil Beds have bite marks the right size to have been inflicted by the powerful jaws of bone-crushing dogs.

Pronounced Ay-**loo**-ro-don, means "cat tooth" in Greek, in reference to the presence of a flange on its upper meat-slicing tooth which modern dogs don't have but cats do.

Remains of this animal have been found in the:
"WATERHOLE" layer (sandstone below ash)

Giant Bone-crushing Dog
Epicyon

Height: about 2 ½ feet (75 cm) at the shoulder and weighed over 200 pounds.

bony crest at top of skull for attachment of huge jaw-closing muscles

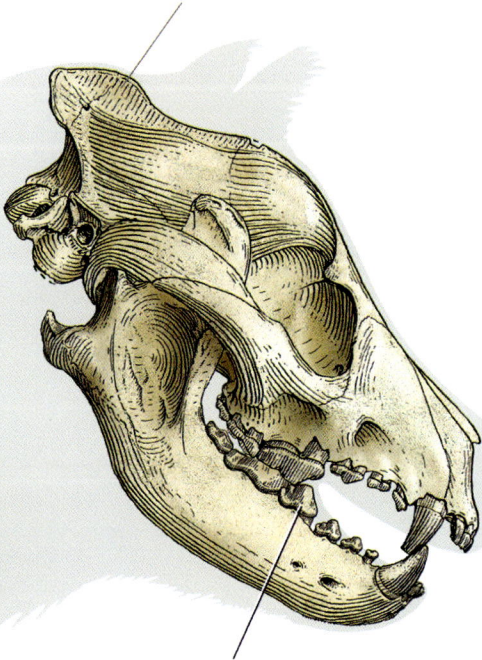

meat slicing tooth tilts backward as in modern hyaenas

The largest members of the dog family (Canidae) that ever lived roamed the North American Great Plains between twelve million and eight million years ago. Many features of their skeletons are similar to those of modern hyaenas in contrast to modern dogs, wolves, and coyotes. Like hyaenas they probably used their huge jaws and teeth to fracture bones. Their enlarged front legs (with flexible forearms) were probably useful in dismembering their prey. Interestingly, giant dogs didn't reach their maximum size until beardogs (Family Amphicyonidae—possibly their direct competitors for carcasses) became extinct about ten million years ago.

The Ashfall species, *Epicyon haydeni,* stood about 2 ½ feet (75 cm) at the shoulder and weighed over 200 pounds; its descendant *Epicyon validus* exceeded 3 feet tall and weighed an estimated 400 pounds.

Literally "top dog," the name refers to the enormous size of the animals.

Remains of this animal have been found in the:
"WATERHOLE" layer (sandstone below ash)

Height: about 1 1/2 feet
(0.45 meters) tall at the shoulder.

Raccoon-like Dog
Cynarctus

Soon after jaws of *Cynarctus* were first discovered in 1901, the animal was classified in the raccoon family by some paleontologists because its wide, blunt molar teeth are reminiscent of modern raccoon teeth. Studies of the ear region, however, confirm that it belongs genetically among the extinct bone-crushing dogs. Unlike its larger bone-crushing relatives, however, *Cynarctus* has comparatively slender jaws and teeth suitable for eating small animals and perhaps fruit and insects as well. The long, sharp-pointed canine fangs may have been useful in dealing with slippery prey like frogs, salamanders, and even fish.

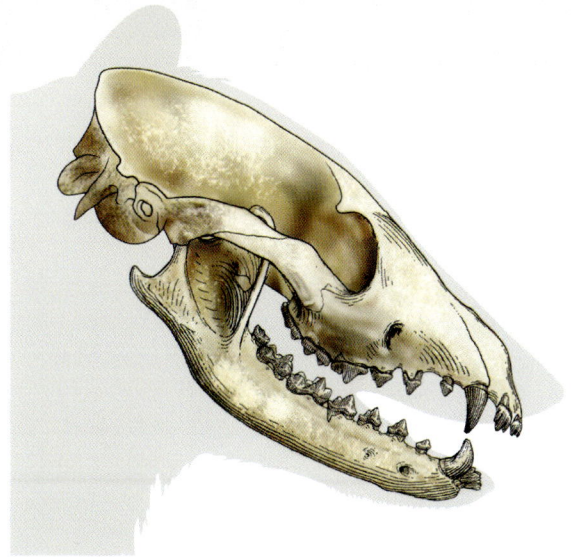

Crowns of the Upper Teeth

Many low cusps for pulverizing small food items.

Small meat-slicing tooth

Sharp fang for piercing.

Comb-like incisors for holding slippery prey.

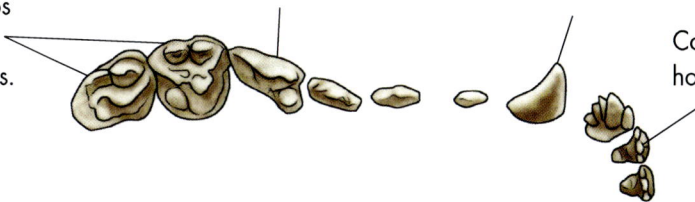

Literally, the name means "dog-bear." It refers to the omnivorous adaptations of the animal's teeth. Pronounced Sine-**ark**-tuss.

Remains of this animal have been found in the:
"DISASTER" layer (volcanic ashbed)
"WATERHOLE" layer (sandstone below ash)

Fruit Dog
Carpocyon

When the wild plums and grapes ripen in late summer, you may catch a lucky glimpse of a coyote stuffing itself on sweet fruit in one of the many fruit thickets here in Ashfall Park. (Check out the plum bushes along the sidewalk just east of the Rhino Barn.) Coyotes, like almost all living species in the dog family, are mostly meat eaters but will take fruit when it is available. In contrast, some extinct dogs like *Carpocyon* (and *Cynarctus*) have teeth and bones better adapted to eating other kinds of food. They lack the specialized meat-slicing teeth and speed-adapted feet that characterize modern dogs.

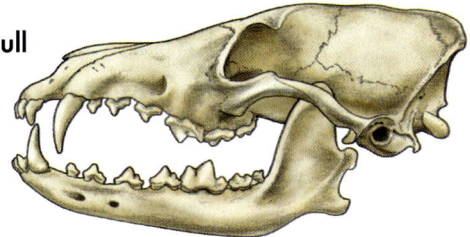

In Greek, *carpos* means "fruit" and *cyon* means "dog." (The blunt molar teeth of this extinct dog suggest that it may have consumed large amounts of fruit.)

Remains of this animal have been found in the:
"WATERHOLE" layer (sandstone below ash)

Fruit dog skull

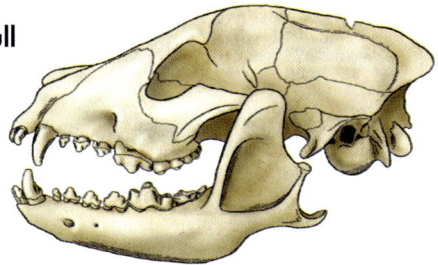

Coyote skull

Back teeth in crown view

molar (crushing) teeth larger in fruit dog

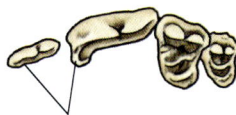

meat slicing teeth long, slender and sharper in coyote

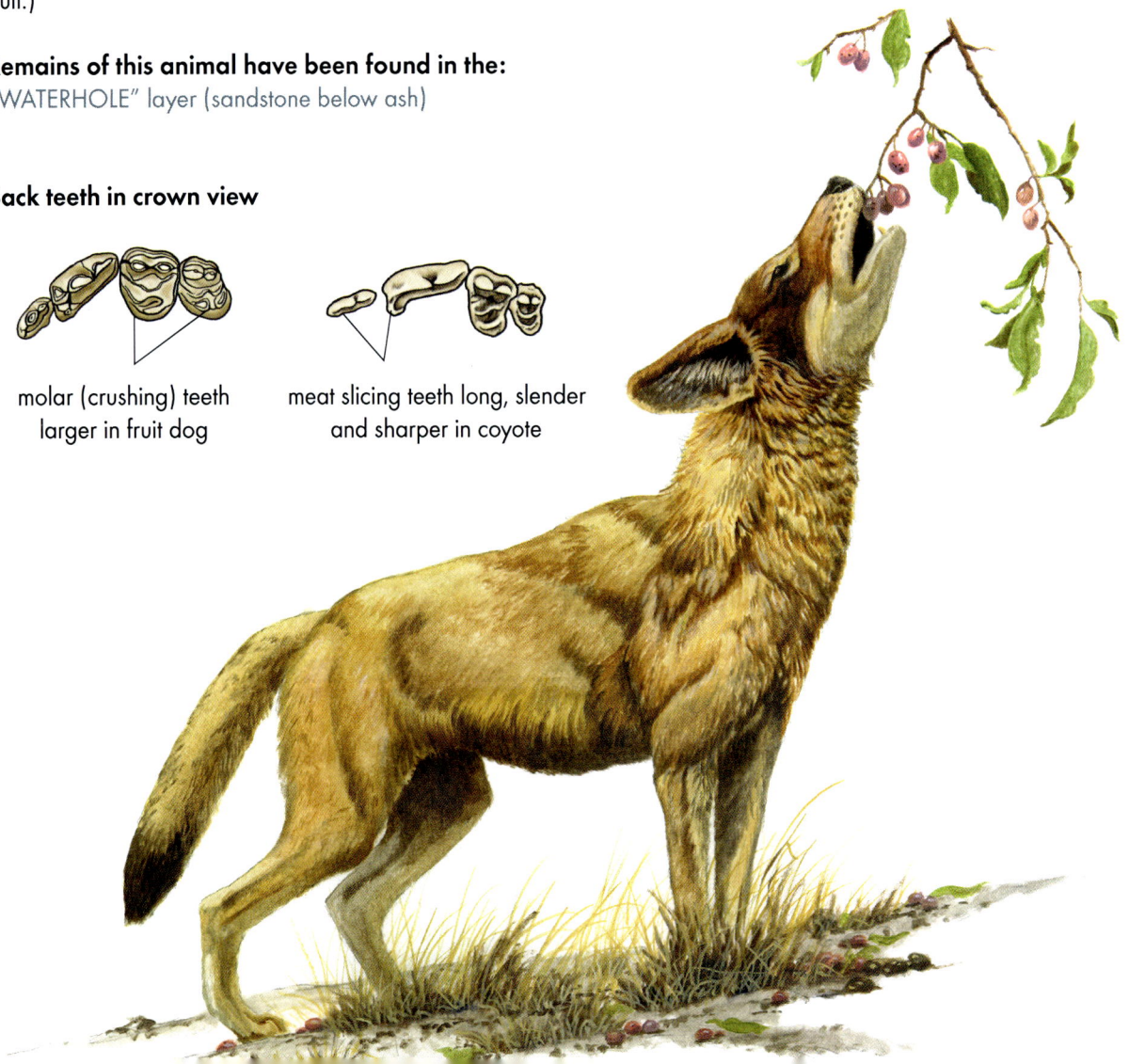

Height: about one foot (0.3 meters) at the shoulder.

New research has shown that all fifty-five living species in the dog family (wolves, coyotes, foxes, jackals, bushdogs, raccoon-dogs, etc. as well as domestic dogs) trace their ancestry back to a small fox-like animal that lived in North America between thirty million and ten million years ago. *Leptocyon* was not only much smaller than contemporary bone-crushing dogs such as *Aelurodon*, *Epicyon*, and *Carpocyon*, but also had proportionately longer lower jaws containing slender, sharp premolar teeth separated by gaps. Like today's red fox, *Leptocyon* was probably adapted to snatching small, fast-moving prey like rodents and birds.

Leptocyon means "slender dog."

Remains of this animal have been found in the:
"DISASTER" layer (volcanic ashbed)
"WATERHOLE" layer (sandstone below ash)

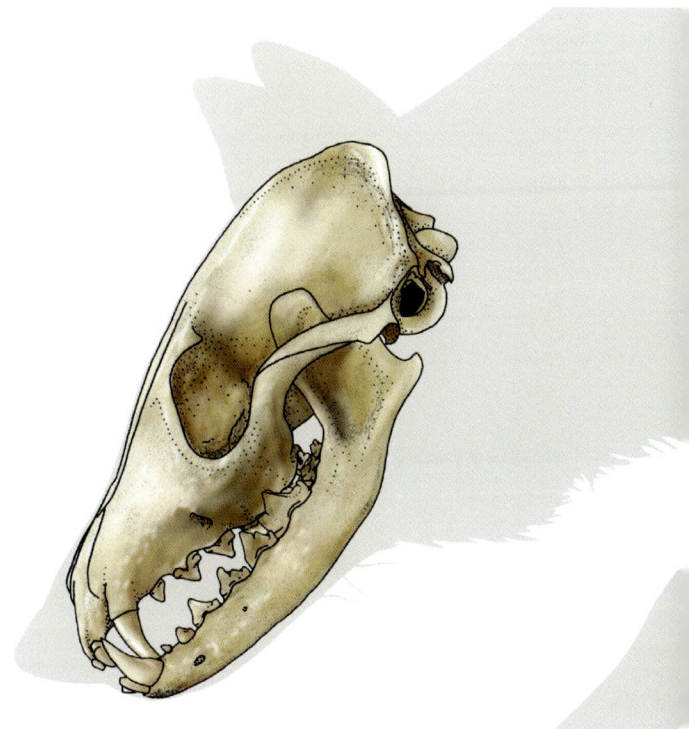

Beardog
Ischyrocyon

Height: about 3 1/2 feet (1.1 meters) at the shoulder.

The largest predators in North America twelve million years ago are classified in an extinct family (the Amphicyonidae) nicknamed the *beardogs*. Despite the name, beardogs were ancestral to neither bears nor dogs, but were on their own evolutionary pathway. Unlike modern bears, they had long heavy tails and meat-slicing teeth in their upper and lower jaws. Unlike dogs, they did not have long legs and foot bones for running but walked flat-footed like bears. They lived in North America until ten million years ago when they were replaced by equally large true bears which immigrated here from Asia.

The diet of beardogs was probably mostly mammals (including carrion) but may have included some insects, fruit, and other plant material.

Ischyros means strong or mighty, *cyon* means dog.

Remains of this animal have been found in the:
"WATERHOLE" layer (sandstone below ash)

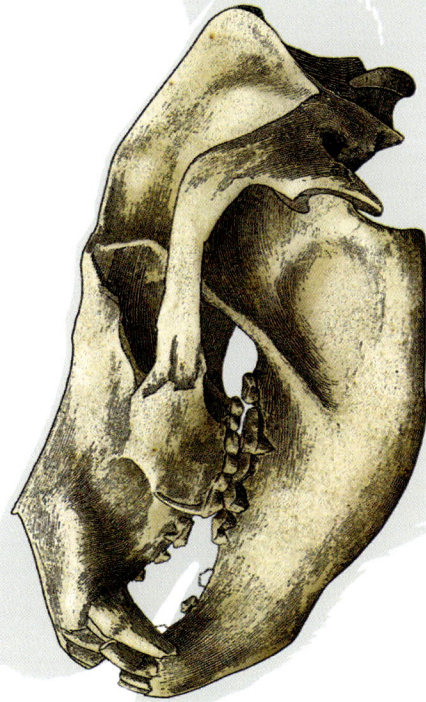

Height: *Barbourofelis whitfordi*, the species expected at Ashfall, probably stood about 2 1/2 feet (0.76 meters) tall at the shoulder. (Geologically later species like *Barbourofelis fricki* reached lion size.)

Thin-sabered "Cat"
Barbourofelis

Barbourofelis was a short-legged saber-toothed carnivore that prowled the Nebraska savannas between twelve million and eight million years ago. It probably couldn't run very well so it attacked from ambush, using its long thin sabers to slash and kill prey. It is near the top of our "most-wanted" list of skeletons yet to be discovered at Ashfall because fierce carnivores are always popular with the public and scientists alike. No articulated skeleton has ever been found anywhere although well-preserved skulls and assorted skeletal parts have been excavated from sites across North America (especially in Nebraska and Florida).

Although not related to true cats, *Barbourofelis* in life probably had some resemblance to a nocturnal ambush cat like the leopard but with shorter legs.

Unlike the bone-crushing dogs, which left abundant evidence of their scavenging activities at Ashfall (broken and gnawed bones, fossilized droppings made of ground bone), *Barbourofelis* didn't have teeth suited to breaking bones. It probably used its razor-sharp teeth to strip meat from its prey and didn't risk damaging its delicate sabers by biting bones.

Literally meaning "Barbour's cat," the name honors Professor E. H. Barbour, a pioneer paleontologist at the University of Nebraska.

Remains of this animal have possibly been found in the:

"WATERHOLE" layer (sandstone below ash)

M. MARCUSON

"Perfect Tusker"
Eubelodon

Height: About as tall as living Asian elephants (males up to 10 feet (3 meters) tall at the shoulder but much more heavily built.)

The largest land animals on Earth twelve million years ago were distant relatives of today's elephants called the *gomphotheres* or four tuskers. The first species to arrive in America, from Asia, between fifteen and fourteen million years ago had very long lower jaws and short tusks in both the upper and lower jaws. Their molar teeth were suited to a diet of soft leaves and fruit. They did not have free-hanging trunks but must have had flexible snouts extending to the tip of the lower jaw. They undoubtedly could drink like a horse or cow by tipping their heads down to get their lips in the water.

By twelve million years ago a descendant of the early four tuskers, *Eubelodon*, had evolved longer legs and a short lower jaw allowing a modern-type trunk to develop.

An enormous upper tusk found just below the ashbed proves that this advanced kind of elephant visited the Ashfall waterhole. Jaws of these animals are visible in the excavation just outside the Rhino Barn.

Eu means "good" or "perfect" and *belodon* means "arrow tooth" or tusk. This elephant was the first in Nebraska (and perhaps the world) to develop large ivory tusks in the upper jaw and a long, free-hanging trunk.

Remains of this animal have been found in the:
"RECOVERY" layer (sandstone above ash)
"WATERHOLE" layer (sandstone below ash)

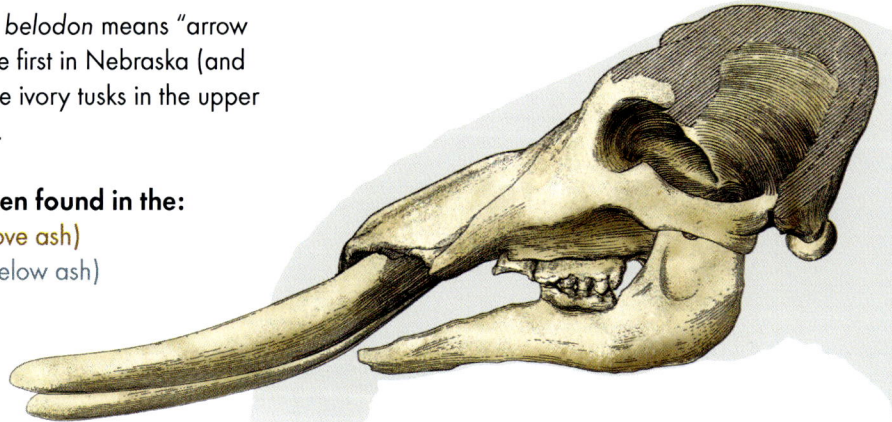

Height: 2 1/2 feet (0.7 meters) at the shoulder.

Peccary
"Prosthennops"

Peccaries are distant relatives of Old World wild boars and domestic pigs. Peccaries originated in North America and spread to South America about three million years ago when the Isthmus of Panama emerged to link North and South America. Today, peccaries are most common in South America (three species) but one species, the javelina, lives as far north as southern Texas and Arizona.

Like other pigs, peccaries are omnivores, devouring a variety of vegetable and animal food–roots, nuts, cactus, and small critters (dead or alive). They can defend themselves with razor-edged tusks in the upper and lower jaws. The tusks are kept sharp by constant grinding against each other.

Canine tusks point straight down unlike wild and domestic boar tusks which curve up. Cheek bones flare outward in some species–giving the animal a "warthog" look. Molar teeth have blunt, rounded cusps (similar to human or bear teeth).

Sthennops means "strong-looking" in reference to the sturdy skull of the specimen first given the name *Prosthennops* in 1904.

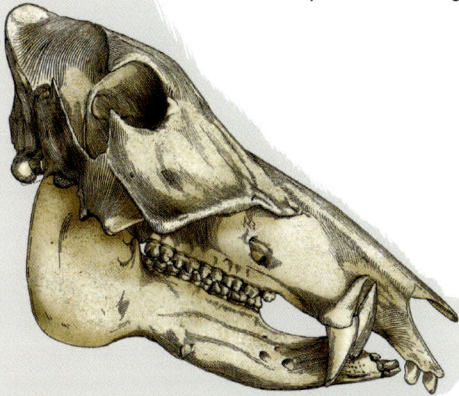

Remains of this animal have been found in the:
"RECOVERY" layer (sandstone above ash)
"WATERHOLE" layer (sandstone below ash)

Round-tailed Beaver
Eucastor

Size: about the size of a muskrat, 2 feet (60 cm) long (including the tail).

In addition to the familiar flat-tailed beavers of today, the fossil record of the beaver family includes two extinct branches: the ancient dry-land beavers which lived like prairie dogs, and the water-loving round-tailed beavers which somewhat resembled today's muskrats and capybaras in anatomy and possible habits. The Ashfall beaver (a single specimen found in 2008 during construction of the Hubbard Rhino Barn) was much smaller than a modern beaver. Some of its descendants, however, were giants—reaching two hundred pounds early in the Ice Age and over five hundred pounds by the late Ice Age.

Pronounced **You**-cast-er, the scientific name comes from Greek: *eu*: "true" or "typical" and *castor*: "beaver."

Remains of this animal have been found in the:
"WATERHOLE" layer (sandstone below ash)

Skull Comparison

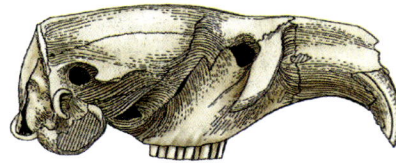

Eucastor skull

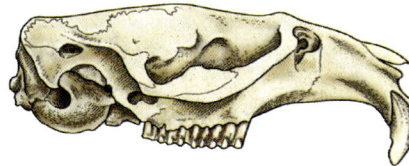

Muskrat skull

Cross-sections of Tail Vertebrae

Round-tailed beaver

Flat-tailed (modern) beaver

Muskrat

Horned Rodent
Mylagaulidae

Size: These rodents were about the size of a prairie dog, 1 foot (0.3 meters) tall when sitting on their haunches.

Most of the dozen or so species of rodents identified in the Ashfall Fossil Beds have close relatives living in the park today: pocket mice, deer mice, gophers, and ground squirrels. But the largest rodent found here has no living counterpart: a species of the bizarre "horned rodent" family that became totally extinct five million years ago.

No full skeletons of these animals have been found at Ashfall but teeth and bones have been collected in all three fossil zones in the park including some in a large fossil burrow preserved in the Rhino Barn. Detailed identification of the Ashfall mylagaulid won't be possible until a good skull is found.

The huge claws and small eye sockets of these animals suggest that they spent most of their time digging underground burrows and eating the roots of plants.

Literally "millstone bucket", the name refers to the single, enormously enlarged grinding tooth dominating both the upper and lower jaws.

Remains of this animal have been found in the:
"RECOVERY" layer (sandstone above ash)
"DISASTER" layer (volcanic ashbed)
"WATERHOLE" layer (sandstone below ash)

Horned rodent skeleton

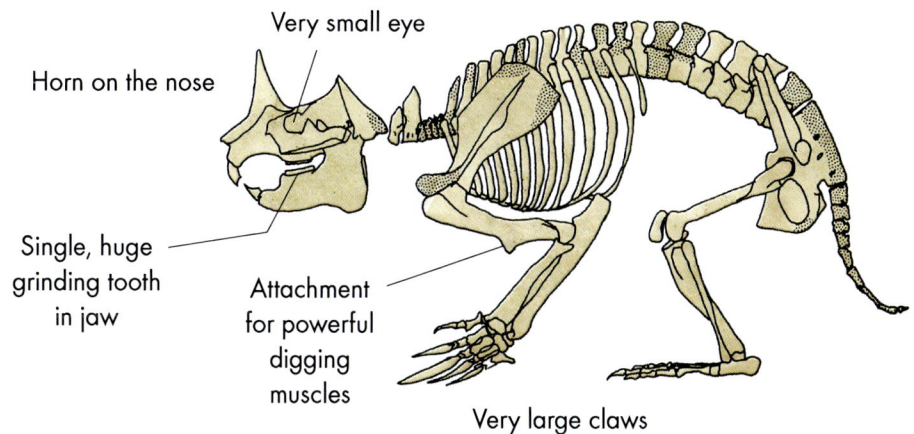

Very small eye

Horn on the nose

Single, huge grinding tooth in jaw

Attachment for powerful digging muscles

Very large claws

Ancestral Kangaroo Rat
Eodipodomys

Size: about a foot (30 cm) long, including the tail.

When most of us hear the word "rat" we think of scary, dirty, scaly-tailed vermin, so it's unfair that one of Nebraska's cleanest, prettiest, and most elegant native animals got saddled with such an ugly nickname. The reason for the "kangaroo" part of the nickname will be obvious to anyone who has driven sandy grassland trails at night and seen the ghost-like little creatures hopping along on their hind legs—just like tiny kangaroos. Along with their close relatives the pocket mice, kangaroo rats evolved entirely in the open grasslands and deserts of western North America. They make their living harvesting seeds of grasses and other plants, and store food for the winter in elaborate underground burrows.

The entire back end of the kangaroo rat skull is made of a thin bubble of bone called the auditory bulla. In modern mammals, an inflated bulla correlates with a well-developed sense of hearing. Modern kangaroo rats have such sensitive ears that they can hear an owl fly!

Fossil skulls of ancient kangaroo rats also had inflated bullae and so probably also had super-acute hearing that would help them avoid predators.

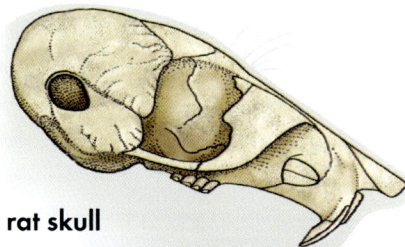

Kangaroo rat skull

Pronounced **Ee**-o-dip-**odd**-o-meez, the name means "ancestral two-footed mouse."

Remains of this animal have been found in the:
"WATERHOLE" layer (sandstone below ash)

Size: The head and body of *Untermannerix* were about 5 inches (12.5 cm) long.

Moonrat
Untermannerix

Hedgehogs and their less-spiny cousins the moonrats don't live in North America today but they lived here between thirty million and ten million years ago when our climate was much milder. "Moonrats"—so called for their nocturnal habits and naked tails—live near water in the tropical forests of southeastern Asia today, eating mostly insects like this crispy cricket—yum!

Named after Mr. and Mrs. George Untermann of Vernal, Utah, who were avid supporters of paleontology; "erix" means "hedgehog."

Remains of this animal have been found in the:
"WATERHOLE" layer (sandstone below ash)

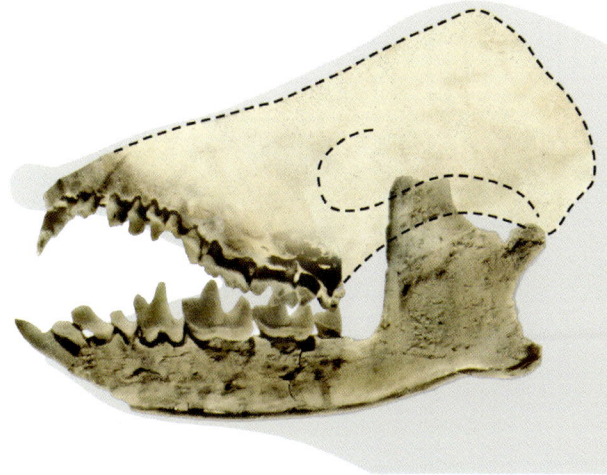

Tiny sharp-pointed teeth are used to pierce the exoskeletons of insects and other small invertebrates.

Secretarybird Mimic
Apatosagittarius terrenus

Height: about 2 feet (0.6 meters) tall.

A partial skeleton of a hawk with very long powerful legs was described as a new species in 1989. When first uncovered, the bird was thought to be closely related to the living Secretarybird of Africa but careful measurements of the toe bones show that in fact a member of the hawk family had evolved into a secretarybird mimic in the ancient Nebraska savanna. Evolutionary *convergence* like this is fairly common: beginning with unrelated ancestors, descendants can become anatomically similar as they adapt to similar environments (in this case the savanna).

Unlike most hawks and eagles, the Secretarybird hunts prey by stalking on foot, then running it down in a zigzag pattern. Prey consists mostly of snakes and other reptiles but small mammals and large insects may also be taken.

Apato means false or deceptive, *sagittarius* is the technical name for the living secretarybird, and *terrenus* means belonging to the ground.

Remains of this animal have been found in the:
"DISASTER" layer (volcanic ashbed)

Foot bones (tarsometatarsus and talons)

Compared to a modern hawk, the foot of the fossil bird is <u>very long</u>, as in the modern Secretarybird.

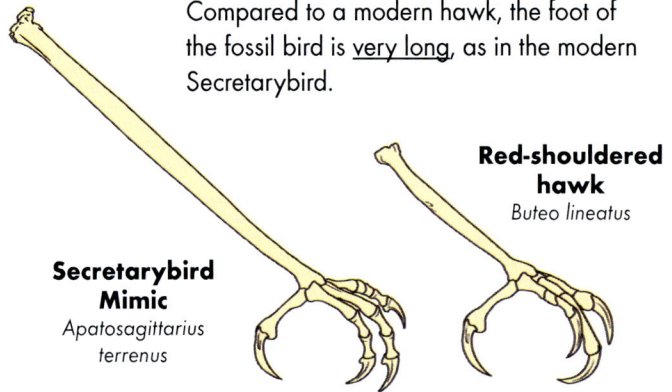

Red-shouldered hawk
Buteo lineatus

Secretarybird Mimic
Apatosagittarius terrenus

The Secretarybird uses its long legs to run down and kill prey–such as snakes–in the savannas of Africa today.

Height: about 2 1/2 feet tall (0.75 meters).

Crowned Crane
Balearica exigua

Perhaps the most surprising discovery in the ashbed was a flock of about forty crowned cranes in the area marked by red flags just west of the Hubbard Rhino Barn. The birds apparently died almost immediately after the initial ashfall. Many were severely trampled but enough skeletons remained intact to establish their identity with the Crowned Crane of Africa. Their most distinctive feature is the sternum (breastbone). In most living cranes, including the Sandhill cranes and whooping cranes that migrate through Nebraska each year, the sternum is hollow and inflated like a bubble of bone into which the trachea (windpipe) makes several loops before continuing into the lungs. This hollow breastbone works as a resonator to amplify the loud, haunting call of cranes in flight. Crowned cranes, which do not migrate, have a sternum like the one in the Nebraska fossil.

Comparison of Breastbones

Living crowned cranes in Africa are classified in the genus *Balearica*. The extinct Ashfall species is named *Balearica exigua*, meaning small, in reference to its size.

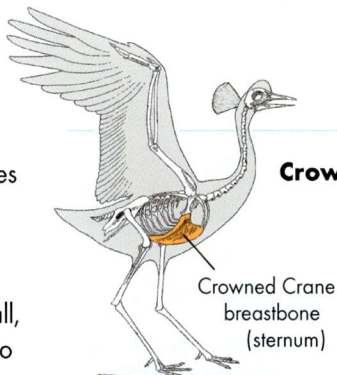

Crowned Crane breastbone (sternum)

Front View

Side View

Sandhill Crane (living)

Sternum inflated into "bubble" that amplifies the birds call

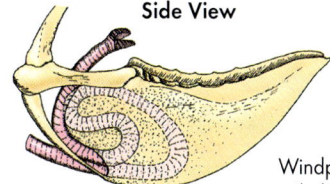

Windpipe enters 'bubble' in sternum and loops around before exiting and continuing to lungs.

Crowned Crane (living and fossil)

Sternum not inflated

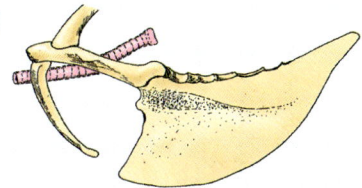

Windpipe (trachea) goes straight to lungs

Remains of this animal have been found in the:
"DISASTER" layer (volcanic ashbed)
"WATERHOLE" layer (sandstone below ash)

Eagle-like scavenger
Anchigyps voorhiesi

Size: about 24 inches (60 cm) tall with a wingspan of 50 inches (127 cm).

This newly-described vulture is an eagle-like scavenger whose closest living relative seems to be the Palm-nut Vulture which ranges over much of tropical Africa today. Scientists concluded that this new species was not closely related to our familiar turkey vultures and condors, for example, but actually forms a link between two groupings of vulture species living today only in Africa and Asia. Maybe it shouldn't surprise us too much since twelve million years ago Nebraska was more like today's East African savanna with a much more diverse grassland ecosystem. The Ashfall waterhole was likely a good place for these birds to hunt and scavenge.

The short, heavy wing bones of the Ashfall vulture suggest that it may not have been such a good glider as the turkey vultures that occasionally visit Ashfall Park nowadays, but may have been capable of strong, flapping flight like an eagle or a hawk.

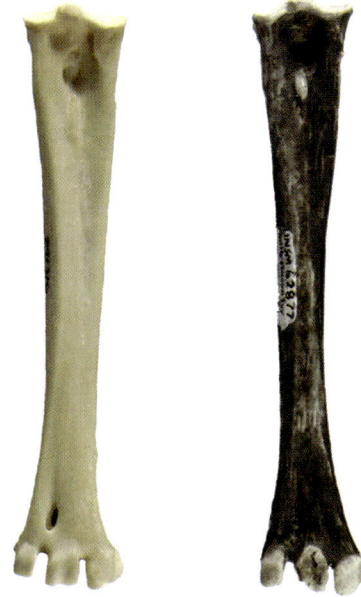

Foot bones of living African Palm-nut Vulture (left) and new Ashfall species (right) showing that they are almost identical in size and shape.

Remains of this animal have been found in the:
"DISASTER" layer (volcanic ashbed)

Giant Tortoise
Hesperotestudo

Size: The largest Ashfall specimen has a shell 4 feet (1.2 meters) long.

Lower jaw

Skull

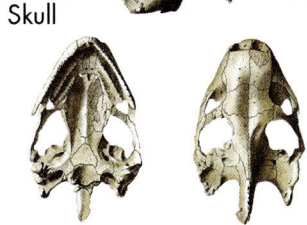

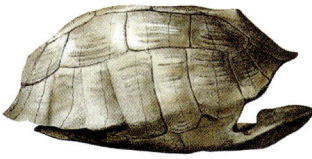

Side view of shell

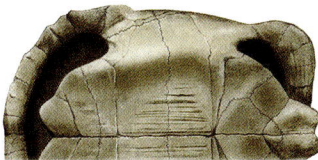

Lower and upper parts of shell

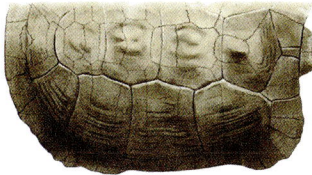

Shells and leg bones of giant land turtles are extremely abundant in the layers of sandstone above and below the volcanic ashbed but only one specimen has been found <u>in</u> the ashbed. This lone individual, "Lonesome George," was buried about three feet above the mass of rhino, camel, and horse skeletons at the bottom of the ashbed. Perhaps because turtles breathe very slowly, he apparently survived much longer (possibly a month or more) than other animals at the waterhole.

Giant tortoises today are strict vegetarians eating a variety of low-growing plants. They cannot survive more than a few hours of freezing temperatures so paleontologists use their fossil remains as evidence for frost-free climate episodes in the geologic past.

As a defense against predators, Ashfall giant tortoises have bony armor embedded in any skin that couldn't be withdrawn inside the shell. Clearly defined growth lines show that Nebraska had well-developed seasons (probably wet and dry) twelve million years ago.

Hespero means western and *Testudo* is the Greek name for the common land tortoise of the Old World.

Remains of this animal have been found in the:
"RECOVERY" layer (sandstone above ash)
"DISASTER" layer (volcanic ashbed)
"WATERHOLE" layer (sandstone below ash)

Sole of hind foot

Fore arm

Small animals that lived in the Ashfall waterhole

Painted Turtle
Chrysemys (new species)

Hundreds of small turtles lived
and died in the Ashfall waterhole
twelve million years ago. Their skeletons
are almost but not quite identical to those
of painted turtles living today in most Nebraska ponds
and lakes. Painted turtles are the most widespread and
adaptable freshwater turtles in North America, able to survive in climatically-variable areas like the Great
Plains where they are subjected to periodic and prolonged drought. They can eat almost any kind of animal
or plant food including dead fish, snails, and duckweed.

Remains of this animal have been found in the:
"RECOVERY" layer (sandstone above ash)
"DISASTER" layer (volcanic ashbed)
"WATERHOLE" layer (sandstone below ash)

Stinkpot
Sternotherus

Stinkpots are little turtles named for the
nasty-smelling liquid they squirt out if disturbed.
They live primarily in shallow water and eat mostly worms and insects. They do not live in Nebraska today but
are found in the eastern and southeastern states of the U.S. Compared to painted turtles, stinkpots don't fare
very well out of water. This may explain why their fossil remains are so rare in the Ashfall waterhole which
apparently dried up frequently. Only one specimen has been found so far at Ashfall.

Remains of this animal have been found in the:
"DISASTER" layer (volcanic ashbed).

Size: *Sternotherus* was up to 5 inches long.
Chrysemys was up to 8 inches long.
Ambystoma was up to 10 inches long.

Axolotl
Ambystoma (larval)

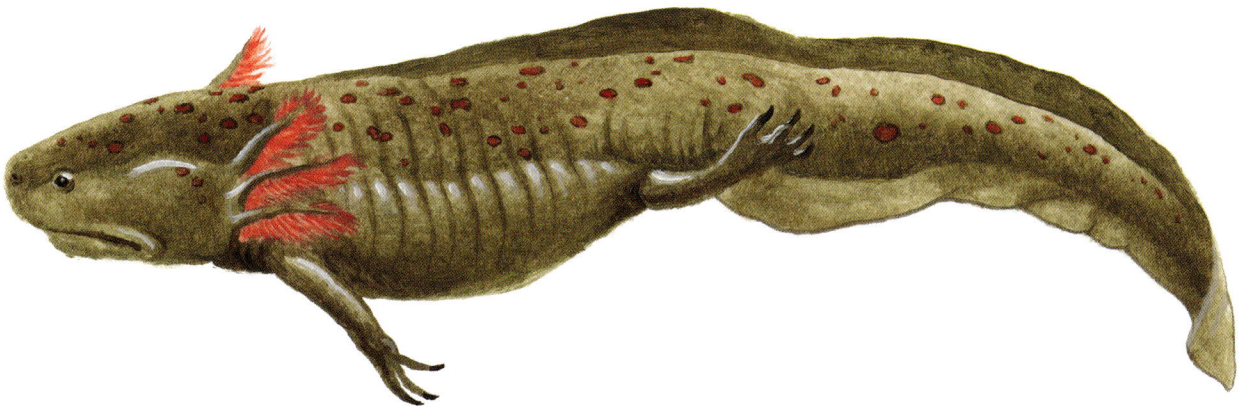

Except for two tiny vertebrae, no fish remains have been collected in the Ashfall waterhole deposits but thousands of bones of small aquatic salamanders have been found below the ashbed. The fossils closely resemble the bones of tiger salamanders that live throughout the Great Plains today. These animals normally lay eggs in water where they develop into larvae with gills. The larvae usually soon grow legs, lose the gills, and spend their adult lives on land. Sometimes, however, the larvae grow to adult size, breed, and lay eggs without ever leaving the water. Such reproductively active but immature-looking animals are called axolotls and their vertebrae can be distinguished from those of normal adults by the presence of a hole for the notochord—a structure present in embryos of all vertebrates but missing in adults that live on land.

Salamander vertebrae in front view

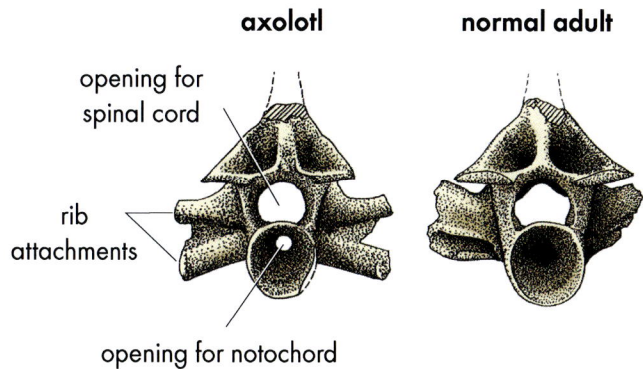

About half of the salamander vertebrae found at Ashfall are perforated for the notochord and therefore presumably had not metamorphosed into normal adults.

Remains of this animal have been found in the:
"WATERHOLE" layer (sandstone below ash)

Why Are So Many Animals of So Many Different Kinds Buried in One Place?

Hundreds of animals crowded into an area no bigger than a small farm pond almost certainly to obtain water after they became sick from breathing in large amounts of volcanic dust.

All available clues indicate that water to drink and cool mud to lie down in lured hundreds of animals to the small area (a few acres) where their fossil remains are so abundant today. The sandstone layer directly under the Ashfall skeleton bed was deposited in a shallow marshy pond that must have existed for hundreds of years before it was obliterated by the volcanic ash storm. Scavenged and trampled fossils of more than sixty species of animals are found in the sandstone, proving that the waterhole had been well known to the local wildlife for many generations before the catastrophe occurred. The Ashfall waterhole may have been a mile or more from the next nearest source of water. Otherwise, how can we explain the crowding that we see in the skeletons of large grazing animals that must have required at least several thousands of acres of grassland to support them while they were alive?

Thirst was probably a major driver of herds of grassland animals in the distant past just as it is today. Death and burial of grazing animals around waterholes is therefore especially common in both ancient and modern grassland habitats. Modern day waterholes are also favorite places for predators to ambush their prey. Fossils in the "waterhole sand" at Ashfall abundantly demonstrate that the local waterhole witnessed plenty of carnage by meat eaters of many shapes and sizes.

At the time the volcanic ash fell, probably early during the rainy season, the deeper parts of the waterhole held several feet of water. The skeletons now being excavated from the ashbed represent animals that probably came to water after they began to become feverish and swollen with Marie's Disease brought on by inhalation of massive amounts of ultrafine ash particles for many days.

Paleontologists have dug many test pits outside the limits of the ancient waterhole and have found no fossils. Conditions for the accumulation and preservation of fossils apparently were present only in the low spot on the ancient landscape that periodically filled up with water (the waterhole) and which collected a much greater thickness of volcanic ash than did surrounding tracts of higher ground.

Visitors interact with paleontologists in the bone bed and lab at Ashfall Fossil Beds.

Photo by Rick Otto, University of Nebraska State Museum

How Does Ashfall Compare with Other Well-Known Fossil Sites?

No two fossil beds are exactly alike. Ashfall is unique in several respects but shares some characteristics with a number of previously discovered deposits in western North America, including a few that are open for public visitation.

Ashfall is the only known locality in the world where whole skeletons of large prehistoric animals are preserved in three-dimensions in volcanic ash. However, as in many other fossil beds dating to the last thirty million years, evidence at Ashfall points to mass death of herds of hoofed animals around a shrinking waterhole.

Within a day's drive of Ashfall Park are three other public sites featuring discoveries of large fossil mammals buried close together; one site is much older than Ashfall and two others much younger.

• Agate Fossil Beds National Monument in westernmost Nebraska preserves the site of one of the greatest fossil discoveries of the early 20th century. Tens of thousands of 20-million-year-old mammal bones closely packed together in an ancient waterhole deposit were quarried by paleontologists from across the U.S. and abroad beginning just over a century ago. Just as at Ashfall, the most common species encountered at Agate was an extinct member of the rhinoceros family, in this case a calf-sized rhino with a pair of horns side by side on the snout. Other species found include an enormous, somewhat horse-like animal with claws instead of hooves and a bison-sized, pig-like scavenger capable of devouring large carcasses—bones and all. Recent excavations at Agate also uncovered ancient burrows containing remains of extinct predators called beardogs. [Fossil burrows have also been found at Ashfall but so far no large skeletons have been found in them.]

• The Hudson-Meng Bison Kill Site in northwestern Nebraska preserves in situ remains of many hundreds of bison. As at Ashfall, a permanent building encloses the excavation site, allowing access to the bone bed for visitors and scientists alike, and protects the fossils from the weather. Although only about 10,000 years old, the bison found at Hudson-Meng average larger than the modern buffalo and are therefore assigned to an extinct ancestral species. Stone spear points have been found at the site but scientists have not yet reached agreement about what killed the bison and to what extent people were involved—did they kill the animals themselves or did they forage among the bodies of victims of a "natural" disaster? The site is still under active study, which makes a visit even more interesting.

• The Hot Springs Mammoth Site in the southern Black Hills of southwestern South Dakota also features a permanent building enclosing a fossil deposit undergoing active excavation but with many fossils left in place for public viewing. Nearly all fossils visible in the bone bed belong to a species of enormous Ice Age elephant called the Columbian Mammoth. Remains of more than fifty-five individual mammoths, nearly all of them young males, judging from their teeth, have been found so far and many more are probably still to be uncovered as excavation continues every summer. Geologic studies show that during the later phases of the most recent Ice Age (twenty to twenty-five thousand years ago) the Hot Springs site was a steep-sided sinkhole with a spring fed pond at the bottom. Mammoths could get in, perhaps attracted by permanent green vegetation or by the warm spring waters, but they couldn't get out. Unlike the rhinos that died together during the volcanic dust storm at Ashfall, the Hot Springs mammoths apparently clambered carefully down into the sinkhole one by one over a period of at least several centuries. Why males only? Modern female elephants and their youngsters live in family groups led by wise old matriarchs who rarely do anything stupid. When young male elephants reach puberty and begin to be obnoxious, they are booted out of the herd and thereafter live mostly solitary lives. They frequently behave cluelessly. It is no surprise then that it was the young male mammoths who couldn't resist the lure of the "Ice Age Hot Tub"—a baited trap at the bottom of the Hot Springs sinkhole.

Although much farther away from Ashfall than the three sites just mentioned, two other public sites featuring massive bone beds should be noted here.

• Rancho la Brea in downtown Los Angeles, California, is another ancient baited trap that attracted Ice Age animals to their deaths and preserved their bones for posterity. Sticky tar was what fatally trapped the animals but it is thought that pools of water sitting on the tar is what originally lured heavy animals like bison, mammoths, camels, horses, and sloths into areas where they became stuck. At that point the trapped animals themselves became the bait and hungry predators such as saber-toothed cats, wolves, bears, and giant carnivorous birds attracted by the dead herbivores entered the treacherous tar pits and were also trapped. Unlike most fossil deposits (Ashfall for example) in which carnivores make up only about one percent of the animals preserved, nine out of ten fossils from the tar pits are from meat eaters. Even though thousands of fossils have been collected from Rancho la Brea, it is not strictly speaking a mass death site like Agate or Ashfall; it is estimated that an entrapment rate of only one animal per month would be sufficient to account for all the bones preserved in the tar because the pits were active traps for many thousands of years, judging from the wide span of radiocarbon dates determined on the bones.

• Dinosaur National Monument in northeastern Utah features a spectacular "logjam" of huge Jurassic dinosaur bones displayed exactly as they were found in a 150-million-year-old river deposit. The fossils are protected by a large building constructed directly over the steeply-tilted layer of sandstone containing the bones. Don't expect to see whole skeletons; the dinosaurs apparently died upstream and had a chance to decompose before their separated bones were washed downstream to a sandbar where they were finally buried.

ASHFALL FOSSIL BEDS:
A Personal Perspective

by Mike Voorhies

What is Ashfall?

Nearly twelve million years ago an enormous blanket of volcanic ash from a distant supervolcano settled over the central Great Plains of western North America. Countless animals and plants perished without a trace on the savanna of what is now eastern Nebraska. But remains of some of them were so deeply and gently buried in ash that they still survive today as exquisitely preserved fossils.

Ashfall Fossil Beds State Historical Park in northeastern Nebraska was developed in 1991 as a place where interested people can get a good close-up look at skeletons of animals killed and preserved by the ash storm and share in the excitement as paleontologists continue to uncover previously unseen fossils every summer. All fossils are being left exactly as they are found; unlike traditional museum fossil exhibits, at Ashfall we have brought the museum to the fossil bed so that visitors can see the actual remains of real prehistoric animals exactly as they exist in nature.

Photo by Mark Harris, University of Nebraska State Museum

Discovery! 1971 "Poison Ivy Quarry"

Many, perhaps most, fossil beds are found accidentally but a few, like Ashfall, were discovered by scientists just doing their job. My wife, Jane, and I, both college geology teachers and researchers, were already in the fourth summer of a long-term field study of the rocks and fossils of northeastern Nebraska when we first visited Melvin Colson's 360-acre farm, now Ashfall Park, in June 1971. We had already explored hundreds of miles of creek banks, roadside ditches, and gullied hillsides on dozens of farms and ranches in northern Antelope and southern Knox counties looking for exposures of bedrock that might contain fossils. We already knew that bones and teeth of extinct animals were fairly common in the area, but were keenly aware that fossils are just curious trinkets unless their exact position in the rock record is known. We were reluctant to attempt any serious fossil collecting until we understood the arrangement, composition, and age of the sedimentary rock layers in the area.

In the Grand Canyon or the Badlands, it's easy to see how the rock layers stack up and it's possible to compile a 'calendar' of the geologic events that took place in those regions. However, in a place like eastern Nebraska where the bedrock is almost totally concealed by row crops, grass, and trees, understanding the sequence of strata is no easy task.

Fortunately for Jane and I, our friend and mentor Morris F. Skinner of Ainsworth had already worked out the basic framework of the rock and fossil record for the central Niobrara River Valley about a hundred miles west of the area we were exploring. Most of the rock layers (formations) studied and named by Skinner extend eastward into the drainage basin of Verdigre Creek north of my hometown of Orchard. One especially interesting rock unit named by Morris is a hard, ledge-forming unit he called the Cap Rock bed, formally called the "Cap Rock Member of the Ash Hollow Formation."

Old-time fossil collectors working along the Niobrara and Snake rivers in Cherry and Brown counties nicknamed the Cap Rock "the fragmental zone" because they often found lenses of abundant but highly fragmentary fossils in it. Skinner reported finding a layer of gray volcanic ash near the bottom of the Cap Rock bed at a number of locations. Jane and I were eager to examine Mr. Colson's property because we suspected that the Cap Rock bed might out-crop there. A newly published detailed topographic map showed that there was a steep northeastward-facing cliff almost a hundred feet high near the southeastern corner of the Colson Farm.

We couldn't see the cliff from the county road but as soon as we had Mr. Colson's permission to access his land, we drove through the gate and soon found it. The cliff was partly covered with vegetation but this was clearly one of the best outcrops of bedrock we had yet seen in four summers of poking around our chosen study area. True to form there was a hard, well-cemented sandstone holding up the top of the cliff, the Cap Rock bed, and beneath that softer sandstones looking just like strata that Morris Skinner had shown us below the Cap Rock bed near Valentine 150 miles to the west. Climbing up to the Cap Rock ledges we found, just as we expected, a thin layer of volcanic ash and some pieces of broken-up rhino, horse, and camel bones jutting from one side of the cliff.

Jane also discovered that the shady cliff was covered with a healthy growth of poison ivy to which she is highly sensitive. She has never gone near the cliff again, leaving me to measure the rocks on the Colson farm by myself. Besides, she had better things to do that summer anyway, taking care of our newborn daughter.

I had planned to spend two or three days measuring the rocks on the Colson Farm. Using the topographic map as a base, I made notes on all the different bedrock exposures I could find, and drew solid lines on the map at the boundary between different rock layers. The Cap Rock bed was especially easy to trace around the farm; it resisted erosion, and its eroded edges tended to be visible as a steep slope, even in areas covered by grass.

I had completed work on the cliff and hiked across to the opposite (northeast) side of the creek valley where there were some gullies at the edge of Mr. Colson's cornfield.

Luckily, the gullies had been flushed clear by recent heavy rains so I could get a good look at the rocks. As expected, the layers looked pretty similar to what I had already measured on the cliff, but one significant difference caught my eye: the volcanic ashbed instead of being a foot or two thick was much thicker. Looking down into the deepest ravine, I could see that the ash was at least six feet thick, maybe more. Holding onto a yucca root, I let myself down the steep side of the gully for a closer look. Sure enough the ashbed was abnormally thick. When I finally shoveled off a smooth surface, my tape measure showed just under ten feet from top to bottom. Morris Skinner had taught me to collect a fresh sack full of ash from the bottom few inches of every new ashbed. Some of the ash samples gathered by Morris during the previous decade had been dated by geologists at the University of California, Berkeley Lab using a relatively new technique called the Potassium-Argon method. This team published a landmark study in 1965 which assigned numerical ages, in years, to many of the major fossil mammal deposits in western North America.

I knew that the laboratory scientists needed crystals, the larger the better, to analyze in order to get reliable dates so I followed out the base of the ashbed with my trowel, looking for the grittiest material I could find. With my nose to the ground, fixated on a futile search for crystals I almost failed to notice something much bigger embedded in the ash—a fossil.

It was a lower jaw of a rhino, not very big, maybe eight or ten inches long. I guessed it was probably a baby and carefully brushed the ash away from several teeth: sure enough they were milk teeth and, what's more, there were a couple of upper milk teeth clenched tight against the lowers, suggesting that the skull was probably there. I'd never seen a fossil in a volcanic ashbed before, so I retrieved the camera from my knapsack and snapped a picture. Obviously, I was curious about how much more of the fossil might still be there, but I still hadn't collected a sample of the ash and hadn't written up my notes on the geology of the gully yet. So I quit looking at the fossil and went back to work. The ash was soft and easy to scrape and brush away and, as hoped, the skull was complete. It was after dark when I got back to our field camp.

Hard though it may be to believe, this was not a "Eureka!" moment for me. I suppose I could fib and say this was a life-changing experience, that with my magical powers I could suddenly envision a vast fossil bed under the hillside. But no, what I actually did see was a fossil bone smaller than my hat with nothing else around it. I checked

Jaw of barrel-bodied rhino calf as spotted by Mike Voorhies in 1971. First evidence of complete skeletons at the Ashfall site.

Photo by Mike Voorhies

both sides of the gully and crawled down the slope looking for evidence that more fossils had washed out of the ashbed before I got there. Nothing.

I've actually been lucky enough to have a few "Eureka" episodes during my fifty plus years of prospecting for fossils, those adrenaline-pumping moments when you just know something important has happened.

At the risk of slowing down this story I'll tell you about two of them. Over the Christmas break from college in 1962, I strutted across a creek bank in Knox County where hundreds of bones were sticking out in plain sight: jaws, teeth, skulls, and other bones of dozens of different kinds of animals were just lying there. I ended up spending the next four years studying the bone bed for my PhD dissertation. A few years later I literally fell into a sandy ravine in Keya Paha County and landed on a pile of petrified elephant, horse, camel, and prongbuck bones and a live bull snake. Within a few minutes I knew that "Bullsnake Quarry" would help unlock a hundred-year-old mystery in American paleontology: exactly where pioneer geologist F.V. Hayden collected the fossils labeled only "sands of the Niobrara River."

Photo by Annie Griffiths/National Geographic Society

Perhaps I should have known the baby rhino jaw in the ashbed was an especially important fossil, but I didn't. Most single fossils I've seen sticking out of the rocks are just that.

The next day I walked the two miles back to the gully carrying a bucket of plaster and burlap intending to collect the fragile specimen or at least protect it from the weather until I had a chance to dig it out safely on a later trip. I soon learned that excavating in volcanic ash is pure joy: it's soft and easy to trowel away but, unlike sand, it doesn't cascade uncontrollably from beneath a fossil leaving it unsupported. It took only a few hours to determine the limits of the skull and enclose it in a tight plaster jacket. As I lifted the skull from its sparkly nest, I got my first good look at what proved to be the atlas vertebra (the first one behind the skull). At this point I began daydreaming: perhaps the whole skeleton was there! For a few seconds, I was seriously tempted to find out by burrowing into the hillside like a badger in hot pursuit of a gopher. But it only took a moment of sober thought to convince me that blindly tunneling back into the hill was a really dumb idea for a number of reasons: 1) there was at least fifteen

With Mr. Colson's permission in 1977, a small field party from the museum explored the site with hand tools and uncovered about a dozen full skeletons of extinct rhinos and horses embedded in volcanic ash. Above, Mike Voorhies sketches an uncovered rhino.

Photo by Annie Griffiths/National Geographic Society

Two of the five horse species found at Ashfall are single-toed, sharing an affinity with today's domestic horse, zebras and the asses.

or twenty feet of ash, sandstone, and sod that would need to be carefully removed from above where the skeleton could be; 2) I didn't know how the skeleton (if it was there at all) was positioned; 3) even if I could expose the skeleton, I didn't have enough plaster to safely jacket a whole rhino and even if I did I couldn't lift it out myself; and 4) intensive fossil-collecting ('quarrying' it's called) was not really my goal for the summer; I wanted to finish up the geologic map.

So, somewhat reluctantly, I packed up my gear and carried the skull back to camp, figuring that I'd eventually return to the site with enough time, people, and equipment to do the job right.

The time finally came in 1977. By then I was working for the University of Nebraska State Museum—previously I'd taught at the University of Georgia and did all my Nebraska fieldwork as a commuter. Our geologic map was finished and we had a chance to investigate some of the dozens of promising fossils found, mostly by Jane, during the summers devoted to mapping.

After getting Mr. Colson's permission to do some shovel work in the gully at the edge of his cornfield, we set to work.

Fortunately Charlie and Karen Messenger, both longtime fossil preparators at the Museum, were on the crew that summer along with three inexperienced but energetic students: Susan Stover, Ken Imig, and Ed Eubanks. Jane kept the field camp running and the diggers well fed.

After clearing the sod from an area about two hundred feet square we scraped down through the ashbed about where we figured the rest of the baby rhino skeleton should be. And sure enough, just like the old song says: "The neck bone connected to the back bone, the back bone connected to the hip bone..." and so on. The whole skeleton was there, fully articulated, just as if in a living animal. Following standard procedure we began to dig a trench around the block of volcanic ash containing the specimen. The Messengers had considerable experience collecting large fossils and were confident that we could collect the skeleton in one large plaster jacket. To do so we needed to carefully clear the ash away from the top and sides of the skeleton and as far beneath it as we could safely dig. It took several days before we had our block ready to cover with strips of burlap soaked in wet plaster of paris. Once the mummy-like "jacket" was completed we had to turn it over and carry it up to our pickup. Luckily volcanic ash is much lighter than sandstone or shale, but our specimen was still a good load for four people. (I was so glad I hadn't tried to collect the specimen by myself six years earlier!)

But the job wasn't over yet. In revealing the baby rhino skeleton we kept running into more and more fossils. Each time our tools made a 'click' we knew we had found another bone and every bone turned out to be articulated with another one. It was unbelievable, even old fossil hands like Charlie, Karen, and I hadn't seen anything like it before. Bone beds in Nebraska's Niobrara River basin are considered very rich by most paleontologists but they consist mostly of separate bones, teeth, jaws, and (occasionally) skulls washed together by ancient rivers. Whole skeletons like the ones we were beginning to uncover just weren't supposed to exist, but here we were, surrounded by dozens of skeletons so close to each other that they almost touched.

We had worked furiously at collecting the skeletons during the few weeks remaining before fall classes started at the University, bringing fieldwork to a close for the year. But with a truckload of plaster jackets to unload at the Museum fossil preparation lab, we were eager to explore the contents of the plaster jackets we had collected.

In retrospect, I date the real "discovery" of the Ashfall Fossil Beds to those weeks during the summer of 1977 when the extraordinary nature of the site first became clear. We showed Mr. Colson what we had found and he kindly donated the fossils to the University. He didn't have to do that; the fossils, like everything else on his land, belonged to him, but proud Nebraskan that he was, he knew that the fossils should belong in the Nebraska State Museum.

(Above) Mike Voorhies carries an excavated rhino skull. (Below) The 1978-79 field crew shows off one week's worth of fossils safely stowed away in plaster jackets.

Photos by Annie Griffiths/National Geographic Society

(Above) The National Geographic excavation in 1978 and 1979 revealed several dozen skeletons from a variety of mammal and bird species that died in close proximity to each other in the ancient waterhole. This photo shows the tight working conditions in the dig site as paleontologists worked to avoid damaging already exposed skeletons.

(Below) Distribution of rhinoceros skeletons in Poison Ivy Quarry, Cap Rock Member, Ash Hollow Formation, Antelope County, Nebraska. Sex is indicated for adults only.

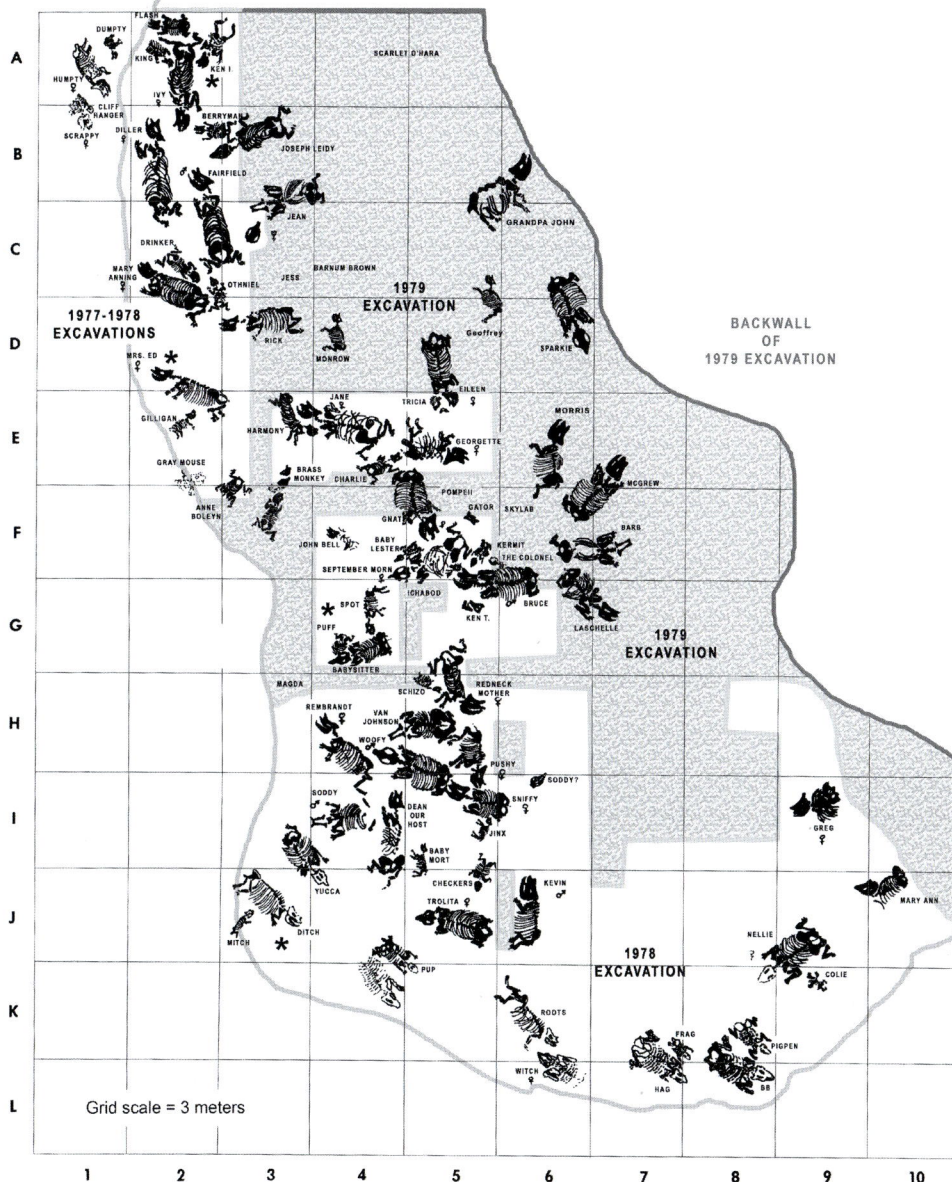

Fossils Had Been Found on the Colson Farm Long Before 1971: Bone fragments in the 1920s and a rhino skull in 1953

The Maple brothers, Forrest and Jerry, grew up in a farmhouse in what is now Ashfall Park (a hiking trail in the park leads down through the abandoned farmyard). The boys remembered picking up fragmentary bones and teeth along the stream that ran through their place. They were not impressed; there's probably not a fifty yard stretch of Verdigre Creek or its myriad tributaries where a sharp-eyed youngster can't find a pocketful of interesting stuff: bones, petrified wood, odd-looking teeth, colored rocks, stone artifacts, spark plugs, and other relics of the near or distant past, all washed together willy-nilly.

The boys didn't keep their finds, but in 1953 a much more important discovery was made by another local youngster. Donald Peterson, the son of a local farmer who had been hired to custom farm a cornfield east of what is now the Ashfall Park Visitor Center, was an avid collector of artifacts and fossils. Taking a break from driving the tractor, he was exploring the steep gullies leading down into the stream valley when he discovered something very unusual: a large fossil skull embedded in a layer of sandstone. Although the specimen was hard and stony ("petrified") it was very brittle. Mr. Peterson wisely decided not to try to dig it out, but instead contacted the University Museum. Soon two experienced paleontological collectors, Lloyd Tanner and Henry Reider, arrived from Lincoln and immediately identified Mr. Peterson's find as a rhino skull. Melvin Colson, owner of the property, agreed to donate the specimen to the Museum, so Tanner and Reider carefully collected it using the tried-and-true method of enclosing it in a sturdy plaster of paris "jacket." They did not report seeing any other fossils in the area at the time and so returned to the Museum with the specimen. The plaster jacket was numbered and the location was listed in the museum's catalog, but no further work was done on the Peterson skull until more than thirty-five years later when exhibits were being prepared for display in the Visitor Center in the soon-to-be-opened Ashfall Park. Newspaper stories about the proposed park naturally prompted the Peterson family to inquire what had happened to the specimen taken to the Museum in 1953.

Unfortunately, I had never heard of the specimen before. The plaster jacket had been sitting on a shelf unopened for more than three decades perhaps because the skull, although well-preserved, did not appear to belong to a new species. Lloyd Tanner was well acquainted with the Museum's extensive fossil rhino collection and he apparently believed that the Peterson specimen was probably so similar to dozens of specimens already in the collection, that the time and expense of detailed work on it could not be justified. Certainly when I was an undergraduate student working for the Museum from 1959 through 1962, I didn't see or hear about the fossil; essentially all research on fossils at that time focused on western Nebraska.

Making the Site into a Public Park: The Idea

The National Geographic Society funded major excavations on the Colson Farm during the summers of 1978-79. People like to see fossils in the ground, especially if they are big and look like real animals. Local people flocked to the Colson Farm to watch the Museum crew excavate the skeleton bed. It became obvious to many of us privileged to participate in those early excavations that folks would drive long distances to witness the kind of work we were doing. Some of us had visited Dinosaur National Monument in Utah, and had been impressed by how well the needs of both scientists and the general public were being met at that location. A large building protected the dinosaur bone bed from the weather, and viewing areas were built so that people could watch the paleontologists at work as they continued to uncover new bones. Could we do something like that here in Nebraska?

Meanwhile, the situation became increasingly awkward for the diggers in the ashbed. Although we would have enjoyed chewing the fat with visitors and showing off what we were finding, we really couldn't do that and accomplish our professional job at the same time. There were hundreds of fossils to carefully remove from the ash once their positions were recorded and mapped. The three summer months available for our work seemed to fly by before we, students and staff alike, had to head back to school. It was hard for us to satisfy the legitimate curiosity of visitors and at the same time protect the delicate fossils from destruction. Each evening before returning to camp we put up signs saying "Please stay out of this area, there's nothing to see but lots to step on."

By the end of the 1979 field season, we had collected all exposed fossils in the ashbed and taken them to the Museum for preparation and study. Personally, I could see no reason for any further work at the site until the fossils already collected could be worked on in the laboratory and studied in some detail. It was eight years before I set foot in "Poison Ivy Quarry" again. By that time, it had gotten a new name and was well on its way to being a public park.

Several seasons of excavation (1987–1990) were required to confirm the presence of fossil skeletons in the ashbed. The grid of "test" trenches were dug through 30,000 square feet of the ashbed. To explore beyond the initial dig, the mapping grid was expanded and trenches were dug along the grid lines. If fossils were found within the trenches, that was proof that there were fossils under the larger area being explored.

Photos by NEBRASKAland Magazine/Nebraska Game and Parks Commission

How Did Ashfall Become a Park?

In my opinion, there probably would be no Ashfall Park if the National Geographic Society had not decided to publish a story about the work they had supported at the site in the 1970s. The eruption of Mount St. Helens in May 1980 had aroused public interest in volcanoes and, accordingly, the January 1981 issue of *National Geographic* magazine was largely devoted to the spectacular blast in Washington State and its devastating consequences. Tucked away in the same issue appeared a story about a volcano-related discovery in Nebraska entitled "*Ancient ashfall creates a Pompeii of prehistoric animals.*" The article featured professional quality photographs of the 1979 fieldwork taken by Annie Griffiths and two dramatic paintings of living and dying rhinos and horses by the outstanding artist Jay Matternes.

Among the Nebraskans surprised and pleased to see the Cornhusker State featured in *National Geographic* were some prominent people who had not only the interest but the financial means to preserve the site and make it available for all to enjoy. Nebraska already had several historical parks, Fort Atkinson, Fort Hartsuff, and Ash Hollow for example, why not a *prehistorical* park celebrating our state's rich heritage of ancient extinct animals? Harold Andersen, long-time editor of the *Omaha World-Herald,* invited me to his office to discuss just such a possibility. Soon he and several fellow board members of the Nebraska Game and Parks Foundation came up with a plan to buy the 360-acre farm and give it to the Game and Parks Commission to develop into a park. The Commission's Director, Gene Mahoney, gave his blessing to the project and in 1986 private donations totaling $385,000 were in-hand to purchase the land, construct an access road and build two buildings, a Visitors Center and an excavation building nicknamed the "Rhino Barn." Major donors included the Ed Owen family, the Burlington Northern Railroad Foundation, and Michael and Gail Yanney. Without the generosity of these visionary philanthropists, Ashfall Park would not be here.

The Park is operated jointly by the Nebraska Game and Parks Commission and the University of Nebraska State Museum. The Commission owns the land and maintains the roads and grounds, while the Museum is responsible for excavation and educational programming. All Park employees, including student interns hired seasonally to help excavate the fossil bed, are University Museum employees.

Photo by Annie Griffiths/National Geographic Society

Photo by Annie Griffiths/National Geographic Society

Photo by NEBRASKAland Magazine/Nebraska Game and Parks Commission

These rhino jaws were collected from the Ashfall site during the summer excavations of 1978–79. Under normal circumstances, there would only be a few isolated jaws available for study from a single site. The mass death from the volcanic ash storm resulted in enough evidence for a population study of extinct species such as the Miocene rhino *Teleoceras* from which these jaws came. This sample of jaws is from all age groups, males and females; providing paleontologists with insight into social structure and lifestyle.

Photo by NEBRASKAland Magazine/ Nebraska Game and Parks Commission

Exciting New Discoveries

Bone-crushing Dog Trackway
Tracks from a bone-crushing dog that walked through the Ashfall waterhole during the volcanic ash storm were discovered in 2014. This is the first recorded example of a twelve million-year-old carnivore trackway.

Raccoon Dog Skull and Skeleton

The first complete skull of *Cynarctus* known to science was discovered during ground preparation for construction of the Hubbard Rhino Barn. Two years later in 2010, a complete skeleton was found inside the new Barn.

Articulated Fossil Snake

The first snake skeleton from the ashbed was discovered within weeks of the *Cynarctus* skull. It is apparently related to the modern constrictor snakes, such as garter, bull, and rat snakes.

Photos by Rick Otto,
University of Nebraska State Museum

References

Beck, D.K., 1995, Hypertrophic pulmonary osteodystrophy recognized in *Teleoceras major* (Mammalia: Rhinoceratidae) from the late Miocene of Nebraska: Geological Society of America Abstracts with Programs, v. 27, no. 3, p. 38.

Boellstorff, J.D., 1976, The succession of late Cenozoic volcanic ashes in the Great Plains: a progress report, Midwest Friends of the Pleistocene, 24th Annual Meeting, Kansas Geological Survey Guidebook Series 1, p. 37–71.

Bonnichsen, B., Leeman, W.P., Honjo, N., McIntosh, W.C., and Godchaux, M.M., 2008, Miocene silicic volcanism in southwestern Idaho: geochronology, geochemistry, and evolution of the central Snake River Plain: Bulletin of Volcanology, v. 70, p. 315–342.

Burchett, R.R., 1986, Geologic bedrock map of Nebraska: Geologic map GMC-1, School of Natural Resources, Conservation and Survey Division, Institute of Agriculture and Natural Resources, University of Nebraska-Lincoln, scale 1:1 000 000, 1 sheet.

Cebula, G.T., Kunk, M.J., Mehnert, H.H., Naeser, C.W., Obradovitch, J.D., and Sutter, J.F., 1986, The Fish Canyon tuff, a potential standard for the 40Ar/39Ar and fission-track dating methods: Conference on geochronology, cosmochronology, and isotope geology, 6th International Meeting, Terra Cognita, v. 6, p. 139–140.

Chappell, W.M., Durham, J.W., and Savage, D.E., 1951, Mold of a rhinoceros in basalt, Lower Grand Coulee, Washington: Geological Society of America Bulletin, v. 62, p. 907–918.

Czaplewski, N.J., Bailey, B.E., and Corner, R.G., 1999, Tertiary bats (Mammalia: Chiroptera) from northern Nebraska: Transactions of the Nebraska Academy of Sciences, v. 25, p. 83–93.

Damuth, J., and Janis, C.M., 2011, On the relationship between hypsodonty and feeding ecology in ungulate mammals, and its utility in palaeoecology: Biological Reviews, v. 86, p. 733–758.

Dawson, M.R., 2008, Lagomorpha, *in* Janis, C.M., Gunnell, G.F., and Uhen, M.D., eds., Evolution of the Tertiary mammals of North America, Volume 2: Small mammals, xenarthrans, and marine mammals: New York, Cambridge University Press, p. 293–310.

Diffendal, R.F., Jr., and Voorhies, M.R., 1994, Geologic framework of the Niobrara River drainage basin and adjacent areas in South Dakota generally east of the 100th meridian west longitude and west of the Missouri River: Nebraska Geological Survey Report of Investigations No. 9, p. 1–13.

Diffendal, R.F., Jr., Voorhies, M.R., Voorhies, E.J., LaGarry, H.E., Timperley, C.L., and Perkins, M.E., 2008, Geologic map of the O'Neill 1°x2° quadrangle, Nebraska with configurations maps of the surfaces of formations: Geologic map GMC-34, School of Natural Resources, Conservation and Survey Division, Institute of Agriculture and Natural Resources, University of Nebraska-Lincoln, scale 1:250 000, 1 sheet, 29 p. text.

Diller, A., 1955, James McKay's journey in Nebraska in 1796: Nebraska History, v. 36, p. 123–128.

Evander, R.L., 1986, Carnivores of the Railway quarries local fauna: Transactions of the Nebraska Academy of Sciences, v. 14, p. 25–34.

Feduccia, A., and Voorhies, M.R., 1989, Miocene hawk converges on secretarybird: Ibis, v. 131, p. 349–354.

Feduccia, A., and Voorhies, M.R., 1992, Crowned cranes (Gruidae: *Balearica*) in the Miocene of Nebraska: Natural History Museum of Los Angeles County Science Series, v. 36, p. 239–248.

Flynn, L.J., and Jacobs, L.L., 2008, Castoroidea, *in* Janis, C.M., Gunnell, G.F., and Uhen, M.D., eds., Evolution of the Tertiary mammals of North America, Volume 2: Small mammals, xenarthrans, and marine mammals: New York, Cambridge University Press, p. 391–405.

Fossilworks, 2013, Mammalian taxa from Devil's Gulch Horse Quarry: http://fossilworks.org/bridge.pl?a=collectionSearch&collection_no=18124 (accessed November 2013)

Gansecki, C.A., Mahood, G.A., and McWilliams, M., 1998, New ages for the climactic eruptions at Yellowstone: single-crystal

[40]Ar/[39]Ar dating identifies contamination: Geology, v. 26, p. 343–346.

Goddard, J., 1970, Age criteria and vital statistics of a black rhinoceros population: East African Wildlife Journal, v. 8, p. 105–121.

Gould, S.J. 1989. Wonderful Life: Burgess Shale and the Nature of History. New York: Norton. p. 61.

Gregory, J.T. 1942. Pliocene vertebrates from Big Spring Canyon, South Dakota. University of California Publications., Bulletin of the Department of Geological Sciences., vol. 26, no. 4, p. 307-446.

Herbel, C.L., 1994, Taphonomy of the "Fragmental layer" Cap Rock Member, Ash Hollow Formation, Ogallala Group (Miocene) in northeastern Nebraska [M.S. thesis]: Lincoln, University of Nebraska, p. 1-100.

Hitchins, P.M., 1978, Age determination of the black rhinoceros (Diceros bicornis Linn.) in Zululand: South African Journal of Wildlife Research, v. 8, p. 71–80.

Holling, H.E., Brodey, R.S., and Boland, H.C., 1961, Pulmonary hypertrophic osteoarthropathy: The Lancet, v. 278 (originally published as v. 2), issue 7215, p. 1269–1274.

Holling, H.E., Danielson, G.K., Hamilton, R.W., Blakemore, W.S,. and Brodey, R.S., 1963, Hypertrophic pulmonary osteoarthropathy: Journal of Thoracic and Cardiovascular Surgery, v. 46, p. 310–321.

Holt County Frontier, Vol. 73, No. 23, October 3, 1953.

Hopkins, S.S.B., 2005, The evolution of fossoriality and the adaptive role of horns in the Mylagaulidae (Mammalia: Rodentia): Proceedings of the Royal Society Biological Sciences B, v. 272, p. 1705–1713.

Hulbert, R.C., Jr., 1993, Taxonomic evolution in North American Neogene horses (Subfamily Equinae): the rise and fall of an adaptive radiation: Paleobiology, v. 19, p. 216–234.

Hunt, R.M. Jr., 1998, Amphicyonidae, in Janis, C.M., Scott, K.M., and Jacobs, L.L., eds., Evolution of the Tertiary mammals of North America, Volume 1: Terrestrial carnivores, ungulates, and ungulatelike mammals: New York, Cambridge University Press, p. 196–221.

Janis, C.M., 1988, An estimation of tooth volume and hypsodonty indices in ungulate mammals, and the correlation of these factors with dietary preferences, in Russell, D.E., Santoro, J–P., and Sigogneau-Russell, D., eds., Teeth revisited: Proceedings of the VIIth International Symposium on Dental Morphology, Paris, 1986, Mémoirs de Musée d'Histoire naturelle Paris, Serie C, v. 53, p. 367–387.

Janis, C.M., 2008, An evolutionary history of browsing and grazing ungulates, in Gordon, I.J. and Prins, H.H.T., eds., The ecology of browsing and grazing: Berlin, Springer-Verlag, p. 21–45.

Joeckel, R.M., and Tucker, S.T., 2002, Exceptional late Miocene rodent burrows, east-central Nebraska: Society of Vertebrate Paleontology Abstracts of Papers, v. 22, supplement to no. 3, p. 72.

Joeckel, R.M., and Tucker, S.T., 2013, Well-preserved latest Miocene (Hemphillian) rodent burrows from the eastern margin of the Great Plains, Greely County, Nebraska: Society for Cenozoic Research (TER-QUA), Lawrence, Kansas, Program and Abstracts, p. 15.

Johnson, F.W., 1936, The status of the name "Valentine" in Tertiary geology and paleontology: American Journal of Science, v. 31, p. 467–475.

Korth, W.W., 1997, Additional rodents (Mammalia) from the Clarendonian (Miocene) of northcentral Nebraska and a review of Clarendonian rodent biostratigraphy of that area: Paludicola, v. 1, p. 97–111.

Korth, W.W., 2000, Review of Miocene (Hemingfordian to Clarendonian) Mylagaulid rodents (Mammalia) from Nebraska. Annals of Carnegie Museum, Vol. 69, Number 4, p. 227-280.

Lander, B., 1998, Oreodontoidea, in Janis, C.M., Scott, K.M., and Jacobs, L.L., eds., Evolution of the Tertiary mammals of North America, Volume 1: Terrestrial carnivores, ungulates, and ungulatelike mammals: New York, Cambridge University Press, p. 402–425.

Lanphere, M.A., and Baadsgard, H., 2001, Precise K–Ar, ^{40}Ar/^{39}Ar, Rb–Sr and U/Pb mineral ages from the 27.5 Ma Fish Canyon Tuff reference standard: Chemical Geology, v. 175, p. 653–671.

Lenehan, T.M., and Fetter, A.W., 1985, Hypertrophic osteopathy *in* Newton, C.D., and Nunamaker, D.M., eds., Textbook of small animal orthopaedics: Philadelphia, J.B. Lippincott Company, p. 603–609.

Lockley, M.G., 1990, How volcanism affects the biostratigraphic record, *in* Lockley, M. G., and Rice, A., eds., Volcanism and fossil biotas: Geological Society of America Special Paper 244, p. 1–12.

MacFadden, B.J., 1992, Fossil horses: systematics, paleobiology and evolution of the Family Equidae: Oxford, Cambridge University Press, p. 1-369.

Mair, T.S., Dyson, S.J., Fraser, J.A., Edwards, G.B., Hillyer, M.H., and Love, S., 1996, Hypertrophic osteopathy (Marie's disease) in Equidae: a review of twenty-four cases: Equine Veterinary Journal, v. 28, p. 256–262.

Mair, T.S., and Tucker, R.L., 2004, Hypertrophic osteopathy (Marie's disease) in horses: Equine Veterinary Education, v. 16, p. 308–311.

Mead, A.J., 1999, Enamel hypoplasia in Miocene rhinoceroses (*Teleoceras*) from Nebraska: evidence of severe physiological stress: Journal of Vertebrate Paleontology, v. 19, p. 393–399.

Mead, A.J., 2000, Sexual dimorphism and paleoecology in *Teleoceras*, a North American Miocene rhinoceros: Paleobiology, v. 26, p. 689–706.

Nasatir, A.P., 1952, Before Lewis and Clark; documents illustrating the history of the Missouri, 1785-1804: St. Louis, St. Louis Historical Documents Foundation, 2 vol.

Otto, R.E., February 2012. Ashfall Fossil Update: *Cynarctus*. The Mammoth, Newsletter of University of Nebraska State Museum, p. 18-19.

Otto, R.E., February/May 2013. Another New Bird Species Recognized From Ashfall. The Mammoth, Newsletter of University of Nebraska State Museum, p. 18.

Osborn, H.F., 1898, A complete skeleton of *Teleoceras fossiger*. Notes upon the growth and sexual characters of this species: Bulletin of the American Museum of Natural History, v. 10, p. 51–59.

Perkins, M.E., Brown, F.H., Nash, W.P., McIntosh, W., and Williams, S.K., 1998, Sequence, age, and source of silicic fallout tuffs in middle to late Miocene basins of the northern Basin and Range province: Geological Society of Amerìca Bullctin, v. 110, p. 344–360.

Perkins, M.E., Nash, W.P., Brown, F.H., and Fleck, R.J., 1995, Fallout tuffs of Trapper Creek, Idaho—a record of Miocene explosive volcanism in the Snake River Plain volcanic province: Geological Society of America Bulletin, v. 107, p. 1484–1506.

Perkins, M.E., and Nash, B.P., 2002, Explosive silicic volcanism of the Yellowstone hotspot: the ashfall tuff record: Geological Society of America Bulletin, v. 114, p. 367–381.

Renne, P.R., Deino, A.L., Walter, R.C., Turrin, B.D., Swisher, C.C., III, Becker, T.A., Curtis, G.H., Sharp, W.D., and Jaouni, A–R., 1994, Intercalibration of astronomical and radioisotopic time: The Journal of Geology, v. 22, p. 783–786.

Renne, P.R., Swisher, C.C., III, Deino, A.L., Karner, D.B., Owens, T.L., and DePaolo, D.J., 1998, Intercalibration of standards, absolute ages and uncertainties in ^{40}Ar/^{39}Ar dating: Chemical Geology, v. 145, p. 117–152.

Rose, W.I., Riley, C.M., and Dartevelle, S., 2003, Sizes and shapes of 10-Ma distal fall pyroclasts in the Ogallala Group, Nebraska: The Journal of Geology, v. 111, p. 115–124.

Scheel, K.E. and Rogers, R.R., 2010, A sedimentological analysis of an unconsolidated air-fall ashbed at Ashfall Fossil Beds, Nebraska. Geological Society of America Abstracts with Programs, Vol. 42, No. 5, p. 632.

Schulte, J.J., 1952, The bedrock geology of Knox County Nebraska, [M.S. thesis]: Lincoln, University of Nebraska, p. 1-98.

Serieyssol, C.A., Slijk, W., Edlund, M.B., and Spaulding, S.A., 2003, Miocene diatoms from the Ashfall Fossil Beds near Royal, Nebraska: North American Diatom Symposium, 17th Meeting, Isla Morada, Florida, Abstracts, p. 37.

Skinner, M.F., and Hibbard, C.W., 1972, Early Pleistocene pre-glacial and glacial rocks and faunas of north-central Nebraska: Bulletin of the American Museum of Natural History, v. 148, p. 1–148.

Skinner, M.F., and Johnson, F.W., 1984, Tertiary stratigraphy and the Frick collection of fossil vertebrates from north-central Nebraska: Bulletin of the American Museum of Natural History, v. 178, p. 215–368.

Skinner, M.F., Skinner, S.M., and Gooris, R.J., 1968, Cenozoic rocks and faunas of Turtle Butte, south-central South Dakota: Bulletin of the American Museum of Natural History, v. 138, p. 215–368.

Skinner, M.F., and Taylor, B.E., 1967, A revision of the geology and paleontology of the Bijou Hills, South Dakota: American Museum Novitates, no. 2300, p. 1–53.

Smith JJ, Turner E, Moller A, Joeckel RM, Otto RE (2018) First U-Pb zircon ages for late Miocene Ashfall Konservat-Lagerstatte and Grove Lake ashes from eastern Great Plains, USA. PLoS ONE 13(11): e0207103. https://doi.org/10.1371/ journal.pone.0207103

Smith, Jon J., Joeckel, R.M., Otto, Rick E., Tucker, Shane T. (2018) Life In The Dead Zone: A Diverse Miocene Ichnofauna Preserved In Volcanic Ash, Ashfall Fossil Beds State Historical Park, Nebraska, USA, Geological Society of America Abstracts with Programs. Vol. 50, No. 6 doi:10.1130/abs/2018AM-318389

Stirton, R.A., and McGrew, P.O., 1935, A preliminary notice on the Miocene and Pliocene mammalian faunas near Valentine, Nebraska: American Journal of Science, v. 29, p. 125–132.

Swinehart, J.B., and Diffendal, R.F., Jr., 1990, Geology of the pre-dune strata, in Flowerday, C.F., and Bleed, A.S., eds., An atlas of the Sand Hills: University of Nebraska Conservation and Survey Division Resource Atlas 5a, p. 29–42.

Tedford, R.H., Albright, L.B., III, Barnosky, A.D., Ferrusquia-Villafranca, I., Hunt, R.M., Jr., Storer, J.E., Swisher, C.C., III, Voorhies, M.R., Webb, S.D., and Whistler, D.P., 2004, Mammalian biochronology of the Arikareean through Hemphillian interval (Late Oligocene through Early Pliocene Epochs) in Woodburne, M.O., ed., Late Cretaceous and Cenozoic mammals of North America, geochronology and biostratigraphy: New York, Columbia University Press, p. 167–231.

Tedford, R.H., Galusha, T., Skinner, M.F., Taylor, B.E., Fields, R.F., Macdonald, J.R., Rensberger, J.M., Webb, S.D., and Whistler, D.P., 1987, Faunal succession and biochronology of the Arikareean through Hemphillian (late Oligocene through earliest Pliocene epochs) in North America, in Woodburne, M.O., ed., Cenozoic mammals of North America, geochronology and biostratigraphy: Berkeley, University of California Press, Berkeley, California, p. 151–210.

Tedford, R.H., Wang, X., and Taylor, B.E., 2009, Phylogenetic systematics of the North American fossil Caninae (Carnivora: Canidae): Bulletin of the American Museum of Natural History, v. 325, p. 1–218.

Thomasson, J.R., 1987, Late Miocene plants from northeastern Nebraska: Journal of Paleontology, v. 61, p. 1065–1079.

Tucker, S.T., Otto, R.E., Joeckel, R.M., and Voorhies, M.R., 2014, The geology and paleontology of Ashfall Fossil Beds, a late Miocene (Clarendonian) mass death assemblage, Antelope County and adjacent Knox County, Nebraska, USA, in Korus, J.T., ed., Geologic Field Trips along the Boundary between the Central Lowlands and Great Plains: 2014 Meeting of the GSA North-Central Section: Geological Society of America Field Guide 36, p. 1–22, doi:10.1130/2014.0036(01).

Voorhies, M.R., 1969, Taphonomy and population dynamics of an early Pliocene vertebrate fauna, Knox County, Nebraska: University of Wyoming Contributions to Geology Special Paper 1, p. 1–69.

Voorhies, M.R., 1971, Paleoclimatic significance of crocodilian remains form the Ogallala Group (upper Tertiary) in northeastern Nebraska: Journal of Paleontology, v. 45, p. 119–121.

Voorhies, M.R., 1973, Early Miocene mammals from northeast Nebraska: University of Wyoming Contributions to Geology, v. 12, p. 1–10.

Voorhies, M.R., 1977, Giant tortoises–fossil evidence of former subtropical climate on the Great Plains: Museum Notes, University of Nebraska State Museum, v. 59, p. 1–4.

Voorhies, M.R., 1985, A Miocene rhinoceros herd buried in volcanic ash: National Geographic Society Research Reports, v. 19, p. 671–688.

Voorhies, M.R., 1990a, Vertebrate paleontology of the proposed Norden Reservoir area, Brown, Cherry, and Keya Paha Counties, Nebraska: Technical Report 82-09, Division of Archeological Research, Denver: U.S. Bureau of Reclamation, p. 1–138, A1–A593.

Voorhies, M.R., 1990b, Vertebrate biostratigraphy of the Ogallala Group in Nebraska, *in* Gustavson, T.C., ed., Geologic framework and regional hydrology: Upper Cenozoic Blackwater Draw and Ogallala Formations, Great Plains: Austin, Bureau of Economic Geology, University of Texas, p. 115–151.

Voorhies, M.R., 1992, Ashfall: life and death at a Nebraska waterhole ten million years ago: Museum Notes, University of Nebraska State Museum, v. 81, p. 1–4.

Voorhies, M.R., 1994, The Cellars of Time-paleontology and archaeology in Nebraska: NEBRASKAland magazine, Nebraska Game and Parks Commission, v. 72, p. 1-162.

Voorhies, M.R. and Corner, R.G. 1995. An Inventory and Evaluation of Vertebrate Paleontology Sites Along the Niobrara/Missouri Scenic Riverways Project Corridors. Prepared for U.S. Dept. of Interior, National Park Service.

Voorhies, M.R. and Otto, R.E. 2002. An Evaluation of Tertiary Age Vertebrate Fossil Sites with Exceptional Preservation in the Lower Niobrara River Basin of Nebraska with Context Assessment for the Great Plains and Other Regions of North America. Prepared for U.S. Dept. of Interior, National Park Service.

Voorhies, M.R., and Stover, S.G., 1978, An articulated fossil skeleton of a pregnant rhinoceros Teleoceras major Hatcher: Proceedings of the Nebraska Academy of Sciences, 88th Annual Meeting, Lincoln, Nebraska, Abstracts, p. 48.

Voorhies, M.R., and Thomasson, J.R., 1979, Fossil grass anthoecia within Miocene rhinoceros skeletons: diet in an extinct species: Science, v. 206, p. 331–333.

Wang, X., Tedford, R.H., and Taylor, B.E., 1999, Phylogenetic systematics of the Borophaginae (Carnivora: Canidae): Bulletin of the American Museum of Natural History, v. 243, p. 1-391.

Webb, S.D., 1969, The Burge and Minnechaduza Clarendonian mammalian faunas of north-central Nebraska: University of California Publications in Geological Sciences, v. 78, p. 1–191.

Webb, S.D., 1983. The rise and fall of the Late Miocene ungulate fauna in North America. In Nitecki, M.H. ed., Coevolution. University of Chicago Press, p. 267-306.

Webb, S.D., 1998, Hornless ruminants, in Janis, C.M., Scott, K.M., and Jacobs, L.L., eds., Evolution of the Tertiary mammals of North America, Volume 1: Terrestrial carnivores, ungulates, and ungulatelike mammals: New York, Cambridge University Press, p. 463–476.

White, J.A., 1991, North American Leporinae (Mammalia: Lagomorpha) from late Miocene (Clarendonian) to latest Pliocene (Blancan): Journal of Vertebrate Paleontology, v. 11, p. 67–89.

Wood, A.E., 1935, Evolution and relationship of the heteromyid rodents, with new forms from the Tertiary of western North America: Annals of the Carnegie Museum, v. 24, p. 73–262.

Zhang, Z., Feduccia, A., and James, H.F., 2012, A late Miocene accipitrid (Aves: Accipitriformes) from Nebraska and its implications for the divergence of Old World vultures: PloS One, v. 7, p. 1–8.

Species List (Fossil Biota) at the Ashfall Site

Taxa	Common Name	In ashbed	Ingested	Sand below ashbed
PLANTS				
Equisetum sp.	Horsetail rush	-	-	X
Carex graceii		-	-	X
Cyperocarpus pulcherrima	Sedges	-	-	X
Cyperocarpus terrestris		-	-	X
Berriochloa communis	Grasses	-	X	X
Berriochloa primaeva		-	X	-
Paleoeriocoma hitchcockii		-	-	X
Juglandicarya sp.	Walnut	-	-	X
Celtis occidentalis	Hackberry	-	-	X
Cryptantha auriculata	Borage	-	-	X
FISH				
Osteichthyes (Gen. et sp. indet.)	Minnow-sized bony fish	-	-	X
AMPHIBIANS				
Ambystoma minshalli	Extinct salamander	-	-	X
Ambystoma tigrinum	Tiger salamander	-	-	X
Bufo valentinensis	Extinct toad	-	-	X
cf. *Lithobates pipiens*	Leopard frog	-	-	X
REPTILES				
Sternotherus odoratus	Stinkpot	-	-	X
Chrysemys n. sp.*	Painted turtle	SK	-	X
Hesperotestudo orthopygia	Giant tortoise	SK	-	X
Sceloporus sp. A	Fence lizard	-	X	-
Sceloporus sp. B	Fence lizard	-	X	-
Cnemidophorus cf. *C. sexlineatus*	Six-lined racerunner	-	X	X
Plestiodon sp.	Skink	-	X	X
Colubrid (Gen. et sp. indet.)	Common constrictor	SK	-	-
Paleoheterodon tiheni	Extinct hognose snake	-	-	X
Ameiseophis robinsoni	Extinct snake	-	-	X
Salvadora paleolineata	Extinct patch-nose snake	-	-	X
Nerodia sp.	Water snake	-	-	X
Thamnophis sp.	Garter snake	-	-	X
Crotaline (Gen. et sp. indet.)	Pit viper	-	-	X

Key: X = present, SK = skeleton

Taxa	Common Name	In ashbed	Ingested	Sand below ashbed
BIRDS				
Balearica exigua*	Extinct crowned crane	SK	-	X
Apatosagittarius terrenus*	False Secretary bird	X	-	X
Anchigyps voorhiesi*	Old World vulture	X	-	-
Rallid (Gen. et sp. indet.)	Rail	X	-	X
Passerine (Gen. et sp. indet.)	Small songbird	-	-	X
MAMMALS				
Ochotonidae				
Gen. et sp. indet.	Pika	-	-	X
Leporidae				
Hypolagus cf. *H. fontinalis*	Small rabbit	-	-	X
Hypolagus sp.	Large rabbit	-	-	X
Mylagaulidae				
Ceratogaulus cf. *C. anecdotus*	Extinct horned rodent	B	-	X
Sciuridae				
Spermophilus (*Otospermophilus*) *cyanocittus*	Small ground squirrel	B	-	X
Ammospermophilus junturensis	Large ground squirrel	-	-	X
Castoridae				
Eucastor sp.	Round-tailed beaver	-	-	X
Cricetidae				
Copemys sp.	Deer mouse	-	X	X
Sigmodontine (N. Gen. et sp.)*	Large mouse	-	-	X
Geomyoidea				
Phelosaccomys hibbardi	Gopher	-	-	X
Heteromyidae				
Eodipodomys sp.	Extinct kangaroo rat	-	-	X
Cupidinimus sp.	Pocket mouse	-	-	X
Mioheteromys cf. *M. amplissimus*	Large pocket mouse	-	-	X
Perognathus cf. *P. minutus*	Small pocket mouse	-	-	X
Canidae				
Aelurodon taxoides	Wolf-sized bone crushing dog	-	-	X
Carpocyon cf. *C. webbi*	Fruit-eating dog	-	-	X
Epicyon cf. *E. haydeni*	Large bone crushing dog	-	-	X
Epicyon cf. *E. saevus*	Small bone crushing dog	-	-	X
Cynarctus cf. *C. voorhiesi*	"Raccoon" dog	SK	-	X
Leptocyon sp.	Fox-sized dog	X	-	X
Amphicyonidae				
Ischyrocyon cf. *I. gidleyi*	Extinct beardog	-	-	X

Key: X = present, SK = skeleton, B = burrow

Taxa	Common Name	In ashbed	Ingested	Sand below ashbed
Mustelidae				
Leptarctus sp.	Koala-like carnivore	-	-	X
Erinaceidae				
Untermannerix copiosus	Moonrat	-	-	X
Metechinus nevadensis	Large hedgehog	-	-	X
Talpidae				
Domninoides n. sp.*	Giant mole	-	-	X
Gen. et sp. indet.	Small mole	X	-	X
Soricidae				
Gen. et sp. indet.	Shrew	-	-	X
Vespertilionidae				
Myotis sp.	Bat	-	-	X
Tayassuidae				
"*Prosthennops*" sp.	Peccary	-	-	X
Merycoidodontidae				
Ustatochoerus skinneri	Oreodont	-	-	X
Camelidae				
Protolabis heterodontus	Llama-like camel	SK	-	X
Procamelus grandis	Ancestral camel	SK	-	X
Megatylopus sp.	Large camel	X	-	X
Aepycamelus sp.	Giraffe-like camel	X	-	X
Gelocidae				
Pseudoceras sp.	Small hornless ruminant	-	-	X
Moschidae				
Longirostromeryx wellsi	Musk deer	SK	-	X
Antilocapridae				
cf. *Proantilocapra* sp.	Pronghorn antelope	-	-	X
Palaeomerycidae				
Cranioceras sp.	Extinct three-horned "deer"	-	-	X
Equidae				
Pseudhipparion gratum	Small three-toed horse	SK	-	X
Neohipparion affine	Slender three-toed horse	SK	-	X
Cormohipparion occidentale	Stout three-toed horse	SK	-	X
? *Protohippus simus*	Slender one-toed horse	SK	-	X
Pliohippus pernix	Stout one-toed horse	SK	-	X
Rhinocerotidae				
Aphelops sp.	Hornless rhino	-	-	X
Teleoceras major	Barrel-bodied rhino	SK	-	X
Gomphotheriidae				
cf. *Eubelodon* sp.	Short-jawed four-tusker	-	-	X

Key: X = present, SK = skeleton

Credits

All animal paintings by Mark Marcuson.

Cover and colorizing of skulls by Angie Fox.

Teleoceras **skull** by Mark Marcuson, University of Nebraska State Museum.

Aphelops **skull** by Edward Drinker Cope and William Diller Matthew, 1915, *Hitherto Unpublished Plates of Tertiary Mammalia and Permian Vertebrata*, Published and distributed with the cooperation of the U.S. Geological Survey by the American Museum of Natural History, Monograph Series No 2.

Neohipparion **skull** by Mark Marcuson, University of Nebraska State Museum.

Cormohipparion **tooth/skull** by Angie Fox, University of Nebraska State Museum.

Pseudhipparion **skull** by Angie Fox, University of Nebraska State Museum.

Pliohippus **skull** by Lindsey Morris Sterling, Henry Fairfield Osborn, 1918, Equidae of the Oligocene, Miocene and Pliocene of North America, iconographic type revision, *Memoirs of the American Museum of Natural History*, new ser., 2(1), plate 30.

Protohippus **skull** by Mark Marcuson, University of Nebraska State Museum (previously unpublished).

Hypohippus **skull** by Lindsey Morris Sterling, 1918, Henry Fairfield Osborn, Equidae of the Oligocene, Miocene and Pliocene of North America, iconographic type revision, *Memoirs of the American Museum of Natural History*, new ser., 2(1), plate 35.

Secretarybird foot courtesy Alan Feduccia and Michael R. Voorhies, 1989, *Ibis*, 131(3).

Secretarybird body courtesy Alan Feduccia, 1980, The Age of Birds, *Harvard University Press*.

Crowned crane sternum modified with permission from Paul Johnsgard, 1983, Cranes of the World, *Indiana University Press*.

Anchigyps voorhiesi **foot bones photo** by Zhang, et al., 2012. Permission of Alan Feduccia.

Ustatochoerus **skull** by Ralph Mefferd, Frick Laboratory, 1941, C. Bertrand Schultz and Charles H. Falkenbach, *Bulletin of the American Museum of Natural History, 79(1), p 66.*

Aepycamelus illustration by William Diller Matthew, 1901, *Memoirs of the American Museum of Natural History*, 1.

Procamelus **skull** by Angie Fox, University of Nebraska State Museum.

Protolabis **skull** Angie Fox, University of Nebraska State Museum.

Cranioceras **skull** by Hazel de Berard, 1937, Childs Frick, Horned Ruminants of North America, *Bulletin of the American Museum of Natural History*, 69, Fig 11.

Longirostromeryx **skull** modified by Michael R. Voorhies. 1937, Childs Frick, Horned Ruminants of North America, *Bulletin of the American Museum of Natural History*, 69, Fig 24.

Proantilocapra **skull** by Hazel de Berard, 1934, Erwin Hinckley Barbour, C. Bertrand Schultz, and Ted Galusha, A new antilocaprid and a new cervid from the late Tertiary of Nebraska. *American Museum Novitates*, no. 734.

Aelurodon **skull** by Mark Marcuson, 1990, M.R. Voorhies, Vertebrate Paleontology of the proposed Norden Reservoir Area, *Division of Archeological Research Technical Report 82-09, University of Nebraska-Lincoln.*

Cynarctus **skull** by Raymond J. Gooris, 1991, courtesy Xiaoming Wang, Richard H. Tedford, Beryl E. Taylor, *Bulletin of the American Museum of Natural History*, 243, Fig 58.

Ischyrocyon **skull** by William Diller Matthew and James Williams Gidley, 1902, A skull of *Dinocyon* from the Miocene of Texas. *Bulletin of the American Museum of Natural History*, 16(11).

Epicyon **skull** by Raymond J. Gooris, 1991, courtesy Xiaoming Wang, Richard H. Tedford, Beryl E. Taylor, *Bulletin of the American Museum of Natural History*, 243, Figs 102g and 109b.

Carpocyon **skull** by Mark Marcuson, University of Nebraska State Museum.

Leptocyon **skull** by Mark Marcuson, University of Nebraska State Museum.

Barbourofelis **skull**, C. Bertrand Schultz, Marian Schultz, and Larry D. Martin, 1970, A new Tribe of Saber-toothed cats (Barbourofelini) from the Pliocene of North America. *Bulletin of the University of Nebraska State Museum*, 9(1).

Eubelodon **skull** by Erwin H. Barbour, 1913, *Nebraska Geological Survey*, 4(11).

Prosthennops **skull**, C. Bertrand Schultz and L. Martin, 1970, *Bulletin of the University of Nebraska State Museum*, 10(1).

Hesperotestudo illustrations by Oliver Perry Hay, 1908, Fossil Turtles of North America, Carnegie Institution of Washington, 75.

Eucastor **skull** by Mark Marcuson, University of Nebraska State Museum (previously unpublished).

Horned rodent skeleton by Mark Marcuson, University of Nebraska State Museum (previously unpublished).

Dipodomys **skull** by Mark Marcuson, University of Nebraska State Museum.

Untermannerix **jaw** modified with permission from Thomas H.V. Rich, 1981, Origin and history of the Erinaceinae and Brachyericinae (Mammalia, Insectivora) in North America, *Bulletin of the American Museum of Natural History*, 171(1).

Salamander vertebrae by Mark Marcuson. M.R. Voorhies, 1990, Vertebrate Paleontology of the proposed Norden Reservoir Area, *Division of Archeological Research Technical Report 82-09, University of Nebraska-Lincoln.*

Excerpts of the Introduction printed with permission of *Natural History Magazine*, "From Waterhole to Rhino Barn" by Sandy S. Mosel, September 2004.

Paleobotanical data from Voorhies and Thomasson (1979) and Thomasson (1987); fish from Herbel (1994); amphibians from Holman (2000, personal commun.); reptiles from Holman (2000, personal commun.); birds from Feduccia and Voorhies (1989, 1992) and Zhang et al. (2012); and mammals from Voorhies (1985, 1990b; personal observations), Herbel (1994), Korth (1997, 2000), Czaplewski et al. (1999), Mead (1999, 2000), and Wang et al. (1999).